Albert Henry Tuttle

The Principles of Histology

Descriptive and Practical

Albert Henry Tuttle

The Principles of Histology
Descriptive and Practical

ISBN/EAN: 9783337405410

Printed in Europe, USA, Canada, Australia, Japan

Cover: Foto ©berggeist007 / pixelio.de

More available books at **www.hansebooks.com**

THE PRINCIPLES OF HISTOLOGY

DESCRIPTIVE AND PRACTICAL:

———

BOOK I.

DESCRIPTIVE HISTOLOGY.

———

BY

ALBERT H. TUTTLE.

———

PUBLISHED BY ANDERSON BROS.
UNIVERSITY OF VIRGINIA.
1898.

PREFACE.

This manual represents an effort to state the most important facts of descriptive histology in a manner adapted to the wants of my own classes of students, both academic and medical. It lays no claim to originality save in the arrangement and mode of presentation, and acknowledgement is here made of the extensive use of the best modern treatises and monographs accessible in its preparation. In the portions dealing with the nervous system I am under special obligations to the writings of Cajal and Van Gehucten among others. I am indebted to Dr. Lyman J. Skeen for frequent and valuable aid in the preparation of the book.

This volume will be followed by a second, now in course of preparation, dealing with Practical Histology.

ALBERT H. TUTTLE.

University of Virginia, May, 1898.

CONTENTS OF BOOK I.

PART I: THE CELL AND THE TISSUES.

PART II: HISTOLOGICAL ANATOMY.

CHAPTER XII.

BOOK I.

DESCRIPTIVE HISTOLOGY.

PART I.

THE CELL AND THE TISSUES.

CHAPTER I

Histology is the science that treats of **tissues**; their structure, their components, their development and modification, and their arrangement to form the organs of the body. We may distinguish between **Animal** and **Vegetable** and between **Normal** and **Pathological Histology**, as well as between **Human** and **Comparative Histology**; expressions which define themselves. As the term Histology is most commonly used in medical literature it signifies the normal histology of man.

In its most limited sense the term Histology is applied to the description of the tissues alone. This description is also sometimes called **General Histology**. The discussion of the arrangement of the tissues to form the organs of the body is distinguished as **Histological Anatomy**; or, (less appropriately) **Physiological Anatomy**.

A **Tissue** may be defined as a mass of similar structural elements (cells or cell-derivatives) having similar properties and functions; together with such substances as are characteristically present between the elements (and are, as a rule, formed by their action), serving to unite them together.

This definition, while true of the great majority of tis-sues, cannot be universally applied in a literal sense. There are some tissues (*e. g.*, adenoid tissue) regularly involving more than one kind of element. Such may with propriety be called **Compound Tissues.**

The combination of tissues gives rise to **organs** of defi-nite structure and function; and, conversely, every organ is ultimately resolvable into its component tissues. There are, however, certain tissue-aggregates which, while themselves entitled to be regarded as organs, sustain a similar relation to the structure of the larger and more complex bodies to which that term is applied as do the tissues themselves. Among such are the smaller blood and lymph vessels, small glands, adenoid masses, etc. The con-sideration of these **compound factors of structure** may properly accompany that of the tissues.

There also occur in the body substances (notably the blood) which, while they lack that coherence properly as-sociated with the idea of a tissue, contain cellular or cor-puscular elements analogous to those forming the basis of tissue-structure. While they can hardly be classed as tissues, their study is clearly within the province of His-tology.

The **cells** (or, more strictly, the **corpuscles**) to which reference has been made, although of exceedingly various form and size, are fundamentally similar in that each is **a mass of** (more or less modified) **protoplasm** provided for a portion if not the whole of its life **with a nucleus.** In some cases the outer layer of the protoplasm gives rise to or is modified to form a more or less well-defined membrane or wall of varying composition, thus forming a true cell; but the constant presence of a cell-wall of definite and approximately uniform composition characteristic of the tissue elements of plants cannot be affirmed of those of animals, the converse being more generally true.

The study of cells as living beings, their internal structure, activities and life-histories, or **Cytology,** is an important and rapidly growing branch of Biology. We are here concerned with them as components of the tissues and with their structure and activities as related thereto. Such discussion thereof as this implies, while it logically precedes an account of the tissues, is on the whole best deferred until some practical familiarity with them and with the elements that compose them has been acquired.

It is sufficient at this time to state that the term **protoplasm** is applied to what has until recently been regarded as one substance, of exceedingly complex chemical composition and apparently devoid of structure; clear or slightly granular in appearance under ordinary powers of the microscope; and viscid or semi-solid in consistency: improved methods and appliances of research have, however, shown that it is neither homogeneous nor structureless. It is the seat of the processes which make up physical life; and it and its products compose the living body, which in its earliest stages consists of a number of apparently similar minute nucleated masses of protoplasm (embryonic cells): the specialization in each of these masses of some one function, and the associated modification in the form of the mass and of its solid products, if any, give rise to tissue-differentiation.

The **nucleus** is a body usually spheroidal in form, both physically and optically denser than the protoplasm by which it is surrounded. It is 'clearly seen, even with ordinary microscopes, to be more complex in structure than the latter: a **nuclear membrane**, a more or less definite **intranuclear network**, and occasional granules, the largest of which, when distinctly spheroidal in form, are known as **nucleoli,** being in many cases readily discernible. During cell-multiplication the nucleus is the seat of important changes, which will be discussed in a subsequent chapter, together with its structure and that of the surrounding protoplasm.

While we may recognize a large number of tissue-elements differing from each other in form and size, as, of course, in function, they can all be included in a very few primary groups, as follows.

Epithelial. those lining or investing a free surface.

Skeletal, those forming an investing, supporting or protective framework for an organ or for the whole body.

Contractile or **Muscular,** those producing by their combined action definite movements.

Irritable or **Nervous,** those acting as reservoirs of energy, or as channels for its discharge, or as receivers and distributors of stimuli.

Reproductive (modified epithelial) elements.

CHAPTER II.

THE EPITHELIA.

Epithelium may be defined as a continuous layer of cells always (a) disposed on a free surface; (b) united by an intercellular cement substance; (c) devoid of blood-vessels, though not necessarily of nerve-terminals.

Epithelium of one or another form is normally present on all free surfaces, save some of those known as synovial; it invests the skin and the mucous membranes with their various diverticula, lines the cavities of the nervous axis, occurs in the organs of special sense, in the cavities of the thyroid and similar bodies, and (as endothelium) lines the serous surfaces and the cavities of the heart and vessels. There are also found in the structure of various solid organs (e. g., the thymus) masses of cells, which, while they no longer line free surfaces, are evidently epithelial in character, comparative and embryological studies demonstrating their epithelial origin. The same may be said of the cells which compose the greater portion of the substance of the liver.

The reproductive elements have already been mentioned as epithelial in origin. They are derived from epithelial layers by processes diametrically opposed in character; those of the male being set free from the layer in which they are formed, those of the female sinking (for a time) into the subjacent tissues. Further consideration of their formation will be deferred until the discussion of the organs in which they occur.

The classification of the epithelia is based on the form, arrangement, or special modification of some or all of the constituent cells.

As regards the form of the cells, epithelia may be

A. Flattened, with flattened nuclei, comprising

1. **Squamous**; flattened or scale-like, with bevelled edges; if with vertical borders, then more properly known as

2. **Pavement**; having the form and disposition of tiles. The terms squamous and pavement are often confounded; they are here used to distinguish clearly recognizable differences of form.

B. Isodiametric, or with nearly equal dimensions in various directions, the nuclei central and spherical or nearly so. Known as

3. **Polyhedral**; also called **Cuboidal**, **Spheroidal**, and, from its most frequent occurrence, **Glandular**; where found on curved surfaces the cells are not unfrequently wedge-shaped or pyramidal through compression.

C. Vertically elongated; the nucleus undergoing a similar change of form, and in some cases situated nearer the base than the free end of the cell. Termed

4. **Columnar**: the various modifications of form may be distinguished as cylindrical, prismatic, club-shaped, etc.

As regards the arrangement of cells, epithelia may be

1. **Simple**; composed of a single layer of pavement, cuboidal, or columnar cells.

2. **Transitional**; composed of a layer a few cells in depth, the constituent cells varying but slightly in form.

3. **Stratified**: several cells deep, and more or less distinctly definable into layers. called **Stratified Squamous** or **Stratified Columnar**, according to the form of the most superficial cells.

There are numerous special modifications of epithelial cells occurring in particular localities, as in the organs of special sense: such will be described in their proper connections. Of more general occurrence are the following:

A. **Ciliated** epithelium; usually columnar, occasionally cuboidal, rarely flattened; the free surface of the cell is beset with hair-like or lash-like prolongations of its protoplasm (cilia), capable of vigorous flexion in one direction.

B. **Goblet-cells**: Columnar (rarely polyhedral) mucus-secreting cells in which the undissolved mucigen accumulates in the distal extremity of the cell, forming a viscid transparent mass, while the protoplasm, together with the contained nucleus, is crowded down to the base of the cell. The extremity of the cell is finally forced off by the escaping mucigen, leaving a chalice-shaped structure.

C. **Prickle-cells**: the polyhedral cells in the deeper portion of stratified squamous epithelium have their surfaces beset with fine immobile processes (prickles) connecting adjacent cells; the latter being separated from each other

by intercellular spaces and channels. By some these processes are regarded as continuations of the protoplasm,
by others of the modified surface of the cell. They will be
discussed more fully in connection with the skin.

The term **Endothelium** is applied to the pavement epithelium lining the vascular, serous and to some extent the
synovial surfaces. Embryological considerations led the
earlier histologists to regard it as quite distinct in origin
from other epithelia; more extended knowledge has shown
this distinction to be doubtful, and some have urged
that the use of the term endothelium should be abandoned
altogether. There are reasons why it appears to be well
to retain the term with the significance above indicated:
as thus applied, it can also be defined as a layer of connective tissue cells on a free surface, a definition that will
receive further consideration in another chapter.

Serous endothelium, when viewed from above, presents
a mosaic of polygons whose various dimensions are nearly
equal, and whose boundary lines are straight and short.
Vascular endothelium is made up of cells elongated in
the direction of the vessel which they line, and tapering at
each extremity. The boundary lines are long and sinuous.

The terms **epithelioid** and (more commonly) **endothelioid** are applied to layers of connective tissue-corpuscles
which resemble epithelium in their regularity, but are not
found upon a free surface. The description of these layers
does not, however, properly belong to the discussion of
the epithelia.

The statements above made concerning the form and arrangement of epithelial cells are purposely quite general in
character: it will readily be understood that within the
groups indicated we may have great variety of detail,
the epithelia of different localities having in almost all
cases their own individual characteristics, by which they
can in many instances be clearly identified, as will be
pointed out from time to time in the discussion of the organs containing them.   It should also be clearly understood that the differences in form described are separated
by no hard and fast lines; flattened, isodiametric, and
elongated cells passing into each other by gradations at
times so slight that it is often difficult if not impossible to
say of a particular cell whether, for instance, it should be
termed columnar or polyhedral; and the continuation of
the same simple layer affording in some instances examples of each in turn of the three fundamental forms.

As regards the arrangement of the cells, epithelia can
never be said to pass into each other so gradually, since a
layer of cells must always be either one cell deep or more
than one cell deep.   The alternative is usually between a
simple epithelium and one composed of more or less distinct strata, the cells on the surface differing markedly
from those at the base: an arrangement transitional between these is usually defined, as stated on a preceding
page.   Practically the term transitional is by most histologists applied only to the epithelium, peculiar alike in structure and function, which lines the urinary tract, in connection with which it will be discussed.

The various epithelia of the body are of exceedingly diverse embryonic origin, as will be pointed out in a subsequent chapter: but however diverse their origin, and however much they may differ in the form and arrangement of the elements of the adult structure, they in each case originate as a single layer of isodiametric cells resting upon the subjacent membrane or its precursor. Their subsequent multiplication, if by cleavage in a vertical plane only, gives rise to a simple layer of polyhedral, or, by lateral pressure, columnar cells: if cleavages occur also in a horizontal direction, or irregularly, a stratified epithelium is the result.

The power of multiplication is probably retained throughout life by all epithelia: by some, however, it may be manifested at times only as a process of repair, or, if normally constant, may proceed but slowly: by others, however, the constant and more or less rapid production of new cells is a part of its chief function. This is notably the case with the stratified squamous epithelia, whose outermost cells are continually being discharged from the surface, either by exfoliation or by their union to form such solid masses as the hairs and the nails.

CHAPTER III.

THE CARTILAGE GROUP.

———

The presence of an intercellular cement-substance has been mentioned as characteristic of the structure of the epithelia: this substance, is, however, always small in quantity, and may or may not be the product of the activity of the cells themselves. Where the epithelial elements give rise to formed products of any considerable extent, they are either deposited in a solid form as a direct investment of the cell itself (e. g., the keratin layer forming the wall of a squamous cell from the outer surface of a stratified epithelium), or are discharged on the free surface as secretions either in a semi-solid form (e. g., the mucigen of the goblet cells), or as a more complete solution (e. g., the products of most glands).

The tissues of the group now to be considered, on the other hand, have it as their chief if not as their sole common characteristic that, however much they may differ in appearance and consistency (from transparent, colorless, almost semi-fluid gelatinous tissue to hard, white, opaque dentine), they consist in every case of cells or **corpuscles**, which, as their chief activity, give rise to a relatively large amount of formed products deposited as an **intercellular matrix**.

The matrix thus formed is at first of slight consistency, and homogeneous in structure. It may become strengthened by subsequent direct modification in density and tenacity; by calcification, lamination, or fibrillation; or by various combinations of these methods. The various changes in the structure of the matrix thus produced, with the concomitant changes in the form, number and arrangement of the corpuscles, give rise to a variety of tissues which have an essential community of function directly associated with their fundamental community of structure. The former is illustrated by the facts of comparative anatomy, which show that these tissues replace each other to a very great extent in different vertebrates; and the latter by the substitutions and adventitious growths that occur abnormally in the human body. As will readily be supposed, they have a common embryonic origin.

The term **Skeletal Tissues** is here applied to the members of this group because of their chief function. They are the framework tissues of the body, investing and penetrating every organ, and supporting and protecting every other tissue. On account of their continuity, and the part they play in binding the organs of the body to each other they are also widely known as the **Connective Tissues**; a term, however, more appropriately used in its original significance, as applied to one of the principal divisions of the group.

As will be readily inferred from what has already been stated, the classification of the skeletal tissues is based in part on differences in the form and disposition of the corpuscles, and in part on differences in the structure of the matrix. It is characteristic of the group as a whole that the corpuscles tend to branch irregularly and to remain or become united by the prolongation of their branches into more or less extensive protoplasmic networks whose nodes are the nucleated bodies of the corpuscles. It is within the meshes of this network that the matrix, whether simple or complex in structure, is deposited.

While, however, the branching and intercommunication of the corpuscles just mentioned is shown by the evidence alike of comparative anatomy, of embryology, and of pathology to be characteristic of each of the skeletal tissues in some animals or in some stages and conditions, we find in the healthy adult body of man and the mammals generally differences in this and associated respects which divide the tissues in question into two principal groups quite sharply distinguished from each other as regards the primary structure of the forms included in each, although in some cases the two types are secondarily intermingled.

In the first, or **Cartilage Group,** the corpuscles, which are always of one kind only, (fixed corpuscles) are usually either spheroidal in form, or, as the result of pressure, polyhedral or flattened: in either case, however, they are simple in outline; and are isolated or at least disconnected: in rare cases they are sparingly branched and connected. The matrix is firm, elastic, primarily homogeneous and finely granular, apparently structureless (though some facts indicate an internal structure not yet clearly demonstrated): whether more or less dense, it is

always permeable by diffusion to the nutrient plasma on
which the corpuscles depend for sustenace, lymph-chan-
nels being absent (or very doubtfully present), and a regular blood-supply wanting, although large masses are sometimes sparingly penetrated by blood-vessels: it is sometimes secondarily reïnforced by the intermingling of fibrous
bundles, or the deposition of lime salts.

According to the extent to which the matrix is developed, and, to the character of its reïnforcement, when this
occurs, the various members of the cartilage group may
be classified as follows:

A. Matrix simple, or not reïnforced by fibres:

1. **Cellular Cartilage**: Matrix very scanty, consisting only of thin layers deposited around the
   corpuscles, which are numerous and relatively
   large: sometimes called **parenchymatous cartilage** from its resemblance to the parenchyma of
   plants: occurs in the embryos of man and many
   vertebrates and in the auricular cartilages of small
   mammals, as well as elsewhere in the permanent
   skeleton of some of the lower vertebrates.

2. **Hyaline Cartilage**: Matrix abundant, though
   varying in quantity, the corpuscles solitary or
   gathered into small groups as the result of recent
   subdivisions: translucent, white or bluish-white in
   color, brittle, firm and elastic: the typical form of
   cartilage. Occurs in the encrusting cartilages of all
   freely movable joints (the corpuscles in them being
   numerous, small and near the surface flattened
   vertically); in the laryngeal cartilages, with two

exceptions to be noted later; in the nasal, costal, tracheal and bronchial cartilages; and in nearly all foetal cartilages.

3. **Calcified Cartilage:** Hyaline cartilage is frequently reïnforced in old age, both in man and in the mammals generally, by the regular deposition in the matrix of nodules of lime salts. This process occurs regularly in some of the lower vertebrates to such an extent as to give rise to a tissue almost bonelike in density and forming the principal framework of the body.

B. Matrix reïnforced by the intermingling with it in smaller or larger proportions of fibrous bundles:

4. **Reticular Cartilage:** Matrix continuous, penetrated irregularly by a network of yellow elastic fibres; the corpuscles relatively large and near together, approaching cellular cartilage in this respect. Occurs where great flexibility and toughness combined with elasticity are called for; in the cartilages of the external ear, in the Eustachian tube, in the epiglottis and in the cartilages of Wrisberg and of Santorini in the larynx. On account of its color and structure this tissue is sometimes spoken of as **yellow fibro-cartilage**, and on account of its physical properties as **elastic cartilage.**

5. **Fibro-Cartilage** proper: Matrix largely replaced by bundles of white fibres; the corpuscles small and few in number, resembling those of hyaline cartilage in appearance. Occurs where great tenacity combined with elasticity and moderate flexi-

bility are needed; in the intervertebral disks, in interarticular masses, at the margins of ball-and-socket joints, in the sacro-iliac articulations, in the symphysis pubis. As distinguished from the preceding it is sometimes termed **white fibro-cartilage.**

Fibro-cartilage may also be described as consisting of masses of interwoven bundles of fibrous tissue with small nodules of hyaline cartilage interspersed sparingly in the meshes. So considered, it may be regarded as a mixture of fibrous and cartilage tissues.

Cartilage always originates as a mass of contiguous spheroidal or polyhedral cells. As development proceeds the cells are seen to be separated by thin layers of a colorless substance, which is formed by the deposition about each cell of a layer of matrix substance known as the **capsule** of the cell; cellular cartilage never proceeds beyond this stage: in the case of hyaline cartilage the matrix substance accumulates between the capsules by external deposition, or else is formed by the gradual transformation and removal of the capsules. Reticular cartilage is always pre-formed as hyaline cartilage, the elastic fibres afterwards appearing in the matrix. In fibro-cartilage the cartilaginous substance and the fibrous tissue are said to appear simultaneously.

Cartilage grows by cell-division, which can without difficulty be seen to have been in progress during life in any good section of hyaline cartilage, the corpuscles being found in groups of two, four or more, so related as to clearly indicate their recent origin: in some instances two cells each with a proper capsule can be found within the capsule of the cell from whose division they were devired. Such interstitial growth doubtless proceeds more rapidly in most cases near the surface than in the deeper portions of the cartilage: it may suffice merely for the constant renewal of the tissue, or may proceed with sufficient rapidity to give rise to actual increase in size. Growth in this sense is believed by some to take place chiefly by apposition: that is by the deposition upon the surface of new cartilage substance.

It is customary to mention in connection with the description of cartilage that the matrix consists chiefly of a substance frequently called **chondrogen**, and said to yield **chondrin** on boiling; the latter is defined as a member of the **gelatin** group of compounds. Gelatin is itself obtained chiefly by boiling the fibrous tissues, which are rich in its antecedent, **collagen**. By some chemists the matrix of cartilage is regarded as also consisting largely of collagen, the so-called chondrin being regarded as only an impure or slightly modified gelatin. The matter is one that has no direct bearing upon the structure of the tissues in question (as far as our present knowledge goes) but it is well for the student to understand what is meant by the terms mentioned.

A cartilage, in the anatomical sense of the word, is an organ: that is to say, a particular part of the body having a definite form and function. As such, its description might with propriety be deferred to the second part of this book: in the case of this, however, as of some other organs consisting chiefly (though not solely) of a single tissue, it will be for various reasons desirable to discuss its structure in connection with that of its prevalent tissue.

A **cartilage**, then, may be defined as **a mass of cartilage tissue** having a definite and regular form. It is, especially where composed of hyaline, calcified, or reticular cartilage, usually covered by the **perichondrium**, a thin fibrous membrane moderately rich in blood vessels, which are the sole or chief source of nutriment for the mass: the elastic fibres of reticular cartilage are continuous with this membrane. A cartilage is always devoid of bloodvessels, though large cartilages are sometimes excavated by spaces of greater or less extent, through which bloodvessels pass, accompanied by lymph-vessels and sometimes by fat, forming what is sometimes termed a "cartilage marrow." It is always devoid of nerves and insensitive to pain.

CHAPTER IV.

THE FIBROUS TISSUES.

In the second of the two groups of skeletal tissues above indicated, the **Fibrous Tissue Group,** in addition to the corpuscles primarily associated with the formation of the tissue and permanently located in it (hence called **fixed corpuscles**), there may be present, in some members of the group at least, characteristic accessory or adventitious corpuscles of various kinds. The fixed corpuscles are always irregular in form, with lamellar or filamentous branches, the latter frequently connecting with similar processes from adjacent cells, thus forming a more or less continuous network. The matrix is always homogeneous, transparent and yielding in the embryonic state, but very early becomes penetrated by fine fibrillae running irregularly in various directions: the fibrillation is in most cases extensive, the matrix finally consisting chiefly of a mass of fibres variously disposed in bundles, in more or less closely felted layers, or in clearly defined laminae. In some cases the fibrillation of the matrix is regularly followed by calcification.

The number of tissues which agree in having the general structure indicated as characteristic of the Fibrous Tissue group is larger than that of all the other tissues of the body put together. Three of them, namely, **corneal tissue, bone tissue** and **dentine,** resemble each other and differ from all the rest in the fact that the fibres formed by the union of the fibrillae are always very minute and are closely felted together to form definite lamellae between which or exterior to which the fixed corpuscles are situated: the first of these is remarkable for its extreme transparency; the other two are normally and extensively calcified, forming tissues of great density and firmness. Their further description will best be deferred to a subsequent chapter.

The remaining members of the group constitute the **Fibrous Tissues** proper or the **Connective Tissues** in the more limited sense in which the term may best be used. Of these one is chiefly if not entirely embryonic, existing in the adult human body only in an extremely modified form. As it is an essential constituent of an important foetal structure, it merits a description as a distinct form of connective tissue: and since it is the precursor of most of the others, its discussion may properly precede their classification and description.

Mucous Tissue (or, as it is also called, **gelatinous tissue**): the matrix is at first homogeneous, transparent and semi-fluid in consistency; it is described as albuminous in composition, with the addition of mucin: fibres very early begin to appear in it, their mode of formation being not yet fully determined: the fixed corpuscles are irregular, branching, connected by their slender processes into a network: in addition there are to be seen here and there in the matrix scattered isolated corpuscles which in the fresh

tissue may be seen to move through the jelly-like substance with an irregular or **amoeboid** motion; these are
the **migratory. corpuscles** or **leucocytes** which, as we
shall see, are characteristic of the connective tissues as a
group. Mucous tissue constitutes an important factor of
the umbilical cord, where it forms, under the name of the
jelly of Wharton, the largest portion of the mass lying between the epithelium upon the surface and the blood vessels
in the centre, up to the fifth month; later the fibrillation
which has already begun about the vessels and near the
surface penetrates the whole mass more and more extensively: but a certain amount of sparingly fibrillated mucous tissue always persists. While mucous tissue is a frequent constituent of the skeletal framework of some of the
lower animals, it is represented in the adult human body
(save as a constituent of morbid growths) by the following
structures, if at all.

The vitreous body (or so-called vitreous humor) of the
eye may best be regarded as a modified form of mucous
tissue. In the embryo it possesses for a time all the characteristics of that structure; but in the adult the matrix
undergoes watery degeneration, and fibres are extremely
rare; fixed corpuscles are altogether wanting, the only
corpuscular elements present being a few leucocytes. The
centre of the intervertebral discs of fibro-cartilage contains a soft and yielding mass sometimes regarded as a
form of mucous tissue: it lacks, however, some of the
essential features of that structure and may best be regarded as the remains of the notochord, the embryonic
precursor of the vertebral column, which, while it differs

in some respects from cellular cartilage, approaches more nearly to it than to mucous tissue as here defined. The pulp of the teeth also consists in part of a modified form of gelatinous tissue, which will be more fully described in connection with those organs.

Passing now to the consideration of the fibrous tissues proper, as defined by the limitations recently indicated, it is important to note at the outset that the fibres which in every case enter so largely into their composition are of two kinds, both of which are present in most of the tissues in question, though 'their proportions may vary exceedingly; the characters of the tissues included in the group being in great measure based upon the proportion of the two kinds of fibres, and the modes of their disposition. These are known respectively as **white fibres** and **yellow** or **elastic fibres.**

The former are exceedingly delicate unbranching filaments of collagen (which, as has been stated, is convertible into gelatin by boiling), rarely more than a micron in diameter, and often much less; they are almost always united into **bundles** of varying size; when so united the fibres have a silky appearance, and tend to assume a characteristic waviness in which all the fibres of the bundle participate in such manner as to retain their almost strictly parallel arrangement: the bundles are again frequently united together in larger aggregates sometimes termed **trabeculae**: white fibres are very tenacious and entirely devoid of elasticity.

The elastic fibres are coarser than the white and more variable in thickness, being from one to six micra in diam-

eter in man, and in some animals as much as fifteen micra:
they are prismatic in form, and under some circumstances
appear to be transversely striated: they branch occasionally, and not unfrequently anastomose: when not on the
stretch they tend to assume large sweeping curves, and
the free ends, which break square across, curl up in a characteristic manner: like the white fibres, they are often associated in bundles. They are, as their name implies, eminently elastic, but are of only moderate tenacity, a bundle
of them being far more easily broken across than a bundle
of white fibres of the same size. They are composed of a
substance known as **elastin**, a complex nitrogenous compound which is not converted into gelatin on boiling. The
elastic fibres are not readily affected by weak acids, as are
the white fibres.

But little is as yet known of the mode of formation of
either kind of fibres. It is by some held that they are in
all cases formed by the transformation of a portion of the
protoplasm of embryonic cells, the remainders of which,
either with or without subsequent increase in size, become
the fixed corpuscles of the tissue in which they are found;
this view, however, is urged not so much on account of
observations directly supporting it as of the conviction on
the part of its most positive adherents that the entire living body consists and must consist only of protoplasm
and the immediate products of protoplasmic changes:
there are, moreover, some facts very difficult of explanation from this standpoint. On the other hand, it is claimed
that fibrillation may and does result from a chemical
and physical change in a part of the homogeneous ground-

substance of the matrix along lines not in actual contact with any mass of protoplasm; although it is freely admitted that the proximity of such masses (the corpuscles) may have a decided influence in initiating and determining such a process. In the case of the elastic fibres, there is good reason for the view (first proposed by Ranvier) that the elastin is deposited in the matrix in the form of small globules, which later fuse together to form fibres: this, if true, will account for the transverse striation already mentioned.

The corpuscles of the fibrous tissues also demand preliminary consideration. It has already been indicated that each tissue is characterized by the presence of a specific form of **fixed corpuscle** peculiar to it; in the case of mucous tissue the presence of **migratory corpuscles**, or, as they are very frequently termed, **leucocytes** has been mentioned as a feature in which this tissue may be taken as a type of the group: and reference has been made to other **accessory** or **adventitious corpuscles**. The fixed corpuscles proper to each tissue may best be described in connection with its definition: the nature and origin of leucocytes will be considered later in connection with the description of the lymph and the blood: the accessory corpuscles now call for discussion. They are often spoken of as modified fixed corpuscles: but while one form is certainly and others are possibly derived from these bodies, there is reason to question whether those of one (if not of more than one) kind are not modified leucocytes: omitting further discussion of their origin (save in one instance) they may be described as follows.

What are known as **fat cells** are formed by the fatty transformation of the protoplasm of certain of the fixed corpuscles of one or more of the fibrous tissues: small droplets of oil at first appear scattered in the cell-body; these become more numerous or larger, finally fusing in one large mass, the nucleus being crowded to one side, and the residual protoplasm forming a thin pellicle or cell-wall.

Under the name of **plasma cells** are included certain corpuscles having elongated and sometimes slightly branching bodies with central oval nuclei, the protoplasm of which contains a large number of small vacuoles (of varying size) which contain a clear fluid probably similar in composition to the lymph or blood-plasma, whence the name of these corpuscles.

Plasma cells were first described by Waldeyer. He included under that term not only vacuolated cells, but also what are now known as **granule cells**: these are usually spheroidal in form and devoid of branches; their protoplasm is highly granular: on account of the marked affinity of the granules for eosin (as well as for many aniline dyes) they are sometimes termed eosinophile cells; the same term has been otherwise applied in connection with the blood, as will be indicated later.

The name of **pigment cells** has been given to connective tissue corpuscles (and to some epithelial cells as well) characterized by the presence in their protoplasm of numerous rounded brown or black granules of a substance termed **melanin**. Pigment cells are of very irregular form, commonly branched, and often exhibiting amoeboid motion when examined in a living condition: the pigment

granules are often so numerous and so closely packed that
the nucleus and other structural features are entirely hid-
den; in some cases, however, they are much less abundant.

The following classification of the fibrous tissues is based
upon the general disposition of the fibres present, their
abundance, the proportionate amount of the two kinds of
fibres, the details of their distribution, and on the kinds
and relative quantities of the associated corpuscles.

A. Fibres varying in abundance, solitary or aggregated
into bundles and trabeculae running irregularly in various
directions, loosely interwoven or more or less closely felted
together.

1. **Areolar Tissue:** called by the older histologists
cellular tissue: found beneath the fibrous layer
of the skin as **subcutaneous,** beneath the mucous
membranes as **submucous,** and the serous membranes as **subserous,** in the interspaces between
adjacent organs as **intermediate,** upon their surfaces as **investing** and forming their internal framework as **penetrating** areolar tissue, it is well nigh
continuous throughout the entire body and merits
in the strictest sense the name of **connective tissue.** The matrix consists in part of a semi-solid
homogeneous ground-substance, penetrated in
every direction by interlacing bundles and trabeculae of white fibres and by elastic fibres either solitary or in bundles: the bundles of fibres may vary
greatly both in their total amount and in the proportion of the two varieties; but they are never
so numerous but that irregular spaces of varying

A. Fibres varying in quantity, loosely interwoven, or felted (*continued*).

size termed **areolae** occur in so great numbers as to be practically continuous; a fact of great importance in the history of dropsical and other effusions. The whole structure is penetrated by blood vessels, lymphatics, and by nerves passing through it: it is colorless or whitish, filmy in texture and of but slight tenacity.

The fixed corpuscles, or areolar tissue corpuscles proper, are quite numerous: they are frequently flattened upon the surface of bundles of fibres or situated in the angles where two or more bundles come together, in either case the processes extending along the bundles and their subdivisions: leucocytes are of frequent occurrence: plasma cells and granule cells less so, except in particular localities. Pigment cells are not common in areolar tissue proper in the human body except in certain places: fat-cells are very common here and there in the areolar tissue of well nourished individuals: where they accumulate in particular localities they give rise to

2. **Adipose Tissue:** this, which may under favorable conditions be formed wherever areolar tissue occurs, but which is most common in connection with the subcutaneous, subserous, and intermediate regions, is in effect little else than areolar tissue in which the fixed corpuscles in particular localities become greatly increased in number, filling

A. Fibres varying in quantity, loosely interwoven or felted (*continued*).

up the areolae, and undergo fatty transformation, giving rise to small fat lobules which are gathered together to form the fat masses visible to the naked eye: the blood supply of these localities is always greatly increased, each lobule having a capillary system of its own, while the fibres between the corpuscles undergo no corresponding increase in number, and in some cases are exceedingly scanty.

3. **Retiform Tissue**, or, as it is also called, **reticular tissue**: this, as its name implies, consists chiefly of a network of fibres, or rather of fibre bundles and trabeculae composed chiefly of what are in all probability most nearly allied to white fibres; true elastic fibres are very sparingly present or are wanting altogether, as is the homogeneous ground substance characteristic of areolar tissue. The fixed corpuscles are flattened, adhering closely to the surfaces of the bundles and trabeculae, and are often so numerous as to form an endothelioid investment. Retiform tissue may perhaps be regarded as a modification in the direction of greater stability of areolar tissue (with which it is often directly continuous), and forms the internal framework of some organs, as well as the basis of the two compound tissues known respectively as **adenoid tissue** and **marrow**. The former of these will be described in connection with the lymphatic system, and the latter in connection with bone.

A. Fibres varying in quantity, loosely interwoven, or felted (*continued*).

4. Fibrous Membrane: this differs from areolar tissue chiefly in the fact that the bundles and trabeculae of fibres, both white and elastic, are far more numerous and closely felted together, obliterating the areolae and leaving small space for the interfascicular ground substance: the fixed corpuscles are quite numerous, but are, as a rule, smaller than those of areolar tissue, and, with their nuclei, generally flattened in the direction of the membrane in which they lie; both plasma and granule cells may be occasionally present, and in some cases pigment cells occur in great numbers: fat cells but rarely.

The principal membranous tracts of the body (*e. g.*, the mucous membranes) which support epithelial or endothelial layers are sometimes more or less clearly divisible into a **stroma**, which makes up by far the greatest part of the layer, and a delicate film situated just beneath the epithelium and known as the **basement membrane**, or **membrana propria**; this is sometimes an endothelioid layer of corpuscles, and in other instances a special condensation of the fibres, either white or elastic. Such differentiation does not occur in those **investing membranes** such as the perichondrium already referred to in connection with cartilages, the similar periosteum of bones, etc., which are not associated with a free surface. **Elastic membranes** consist

A. Fibres varying in quantity, loosely interwoven, or felted (*continued*).

chiefly or entirely of elastic fibres, or, in some cases, of continuous layers of elastin.

5. **Fenestrated membrane**: this form of membrane is produced where, in an otherwise normally formed fibrous membrane, there are at intervals of greater or less extent neither fibre bundles nor ground-substance, thus leaving rounded openings of various size and frequency, as in the omentum of man and of numerous mammals. Fenestrated membranes may also consist chiefly, if not entirely, of elastic tissue, in which the fibrous structure at times disappears in great measure, as in the fenestrated elastic membranes of the blood vessels.

B. Fibres exceedingly abundant, one or the other kind predominating, aggregated into smaller and larger bundles, which are in a generally way disposed parallelwise: ground-substance very scanty. Fixed corpuscles few, flattened between the bundles of fibres.

6. **Tendon tissue** or **white fibrous tissue**: as areolar tissue merits in the strictest sense the name of connective tissue, so this, above all others, merits that of **fibrous tissue**. It consists almost exclusively of white fibres: these are united together in small bundles by a small quantity of ground-substance, each being covered by an endothelioid layer of flattened corpuscles: the smaller bundles are gathered together with occasional anastomo-

B. Fibres abundant, chiefly of one kind, in parallel bundles (*continued*).

ses into larger ones of varying size, which are separated by interstitial areolar tissue continuous with the investing sheath of the whole mass: elastic fibres are very sparingly present, chiefly in the penetrating areolar tissue. The characteristic fixed corpuscles or **tendon cells** are generally found in rows which occupy the spaces between two or three contiguous bundles, their branches taking the form of interfascicular lamellae: leucocytes are rarely present, and other forms of corpuscles are wanting. Blood vessels and lymphatics, and, in some cases at least, nerve fibres follow the interstitial areolar tissue. White fibrous tissue is found in tendons and ligaments and also in tendinous aponeuroses and fasciae, which form a transition to fascicular investing membranes.

7. **Elastic tissue**: this, as its name implies, is composed chiefly of elastic or yellow fibres, which are arranged in bundles of varying size and complexity, the interstitial connective tissue penetrating not only the larger but also the smaller bundles, and in some cases separating individual fibres: small bundles of white fibres run in the connective tissue in various directions, and in some cases at least there are well defined bundles arranged in groups as in tendon tissue and running throughout the whole length of the structure. Elastic tissue is generally regarded as differing from all other

B, Fibres abundant, chiefly of one kind, in parallel bundles (*continued*).

> forms of skeletal tissue in having no characteristic
> fixed corpuscles, none being found in it that can
> with certainty be regarded as such : some authors
> have described in this connection flattened corpus-
> cles found scattered in the ground-substance be-
> tween the fibres and in close apposition with the
> latter; it is not, however, certain that these do
> not more properly belong with the interstitial con-
> nective tissue.   The term elastic tissue is by some
> histologists applied alike to the chief constituent
> of elastic ligaments (in which sense it is here de-
> fined), and to that of elastic membranes.

We may now with profit revert to the definition of endothelium given in a previous chapter, which described it as
a layer of connective tissue corpuscles on a free surface.
We have seen that the association of corpuscles with
fibrous structures is characteristic of the connective tissues: embryological evidence shows that the serous and
vascular cavities are alike formed by cleavages or excavations of the mass of embryonic cells from which the skeletal tissues are derived : the tendency of connective-tissue
corpuscles to arrange themselves in layers (properly
termed endothelioid) has been mentioned in connection
with such structures as basement membranes, the investing
layers of small tendons and tendon-bundles, etc.: and, as
we shall see later, the direct continuity between endothelioid cells and connective tissue corpuscles can be observed

at the free extremities of lymphatics and at the margins
of synovial surfaces: it can, therefore, be readily understood how connective tissue corpuscles should form a continuous layer under such exceptionally favorable conditions as those found on the free surface of a membrane.
Some characteristic activities of the corpuscles found upon
serous surfaces will be described in connection with the
lymph and the blood.

We know at present but little of the duration of any of
the fibrous tissues, or, indeed, of the skeletal tissues in
general. Consisting largely of the constituents of the
matrix, which some histologists regard as formed and in
a certain sense not-living substances, it has been held that,
once established, they may endure as long as the body
lasts. It is possible that this is the case, to some extent
at least, and that the tendon fibres, for example, of our
old age are the identical tendon fibres of our childhood:
on the other hand, the tendons of the child certainly increase both in thickness and in length toward manhood;
and the mechanism by which interstitial increase takes
place is certainly adequate for interstitial replacement as
well; we have, however, at present no certain evidence of
any such mechanism of absorption or removal of worn-
out fibres (if such there be) as a method of replacement
would imply. Whether the matrix of a fibrous tissue is to
be regarded as living or not depends entirely on what we
mean by a word for which no generally accepted definition has yet been given: it is certain, however, that under
certain conditions of defective nutrition such structures

undergo marked changes to which the name of death is certainly not inappropriate.

These changes are generally regarded as due primarily to the death of the corpuscles: but this merely shifts the problem. It is difficult to conceive that any single nucleated mass of protoplasm should retain its powers unabated for half a century or more; but if it may, what are its probable activities in the case, for example, of a tendon cell? If, on the contrary (as seems more probable), it is constantly renewed, what are the activities of the successive generations? The phenomena of tissue-repair in the case of injuries throw some light on the problem, but it is, on the whole, at present unsolved. It is desirable, however for the student to know of its existence.

Before dismissing the discussion of the fibrous tissues it is proper briefly to mention a form of tissue at first supposed to be a member of this group, and described as a modification of retiform tissue: subsequent investigations have shown, however, that it differs in important respects from any of the true connective tissues in structure, and is of widely differing embryonic origin. It is the sustentacular tissue of the brain and spinal cord, and is known at present by the name of **neuroglia**. It is composed entirely of branching corpuscles and their fibrillar processes, known as **glia-cells**: their full description can best be given in connection with the nervous tissues with which they are associated.

CHAPTER V.

THE LAMELLATED TISSUES.

As was stated in the preceding chapter, there are three members of the skeletal tissue group (namely, corneal tissue, bone tissue and dentine), which, like most of the fibrous tissues described in that chapter, are characterized by extensive fibrillation of the matrix, the fibrillae being even more closely intermingled than are those of ordinary membranes: they differ from these structures, however, as already indicated, in the facts that the fibrillae, which are always exceedingly fine and resemble most nearly those composing white fibres, are never aggregated into the bundles and trabeculae which are interwoven to form membranes, but are felted together to form dense layers, having in each case the characteristic fixed corpuscles situated between them or exterior to them.

These lamellar structures have evidently much in common with and are originally derived from the modification of membranes, as the facts of comparative anatomy and embryology plainly demonstrate (two of them being clearly dermal in origin in man and the lower animals alike): the characters above stated, are, nevertheless, of such importance as to warrant their separate consideration. The name at the head of this chapter is, therefore,

proposed for the group: but it should always be borne
clearly in mind that they are more nearly related to the
fibrous tissue group than are either of them to the cartilages: this is the more important because of the fact
that one of them, osseous tissue, largely replaces cartilage
in the formation of most of the bones of the body, thus
giving rise to the impression that these two tissues are
closely allied, and that the latter is in some way trans-
formable into the former, an error which leads to much
unnecessary confusion.

The formation of **dentine**, the most peculiar member of
the group, is so intimately related to that of the other tissues of the teeth, (of which it forms the largest part), and
especially with that of the pulp contained in the tooth-
cavity that it cannot well be discussed apart from these
associated structures: they will be considered together
when the teeth are described in connection with the other
organs of the region in which they occur.

Since **corneal tissue** is found only in the structure from
which it derives its name, its consideration might in like
manner with propriety be deferred until the description of
the eye. There are, however, points of resemblance between it and bone tissue which make its study desirable
as a preliminary to that of the latter, particularly as the
absence of calcification renders far easier the recognition
of important details.

The transparent lamellae which form almost its entire
bulk are quite uniform in thickness throughout the entire
cornea: they are composed of white fibres running parallel

to each other in each lamella, so disposed that those of one lamella cross those of the next at right angles, or nearly so, in the centre of the cornea; toward the margins they cross at varying degrees of obliquity: the lamellae near the outer surface of the cornea are traversed obliquely by occasional bundles of fibres, which thus unite them together. Adjacent lamellae are separated here and there by shallow lens-shaped spaces which occur frequently, but at irregular intervals; these may be designated as **lacunae:** those lying between the same two lamellae are connected with each other by numerous branching channels, which may with equal propriety be termed **canaliculi:** there is thus formed a continuous system of canals and spaces across the entire cornea between each two lamellae, the lacunae of one such system having no definite relation in position or otherwise to those of the next.

In the lacunae lie the **corneal corpuscles,** apparently adhering to the surface of one or the other of the two adjacent lamellae. They are flattened, with flattened nuclei, and irregular in outline; they branch freely, the branches extending into the canaliculi and in many cases connecting with those from adjacent corpuscles, thus forming a protoplasmic network, which is possibly coextensive with the canal-system, but does not completely fill it. The remainder of the space forms a means of distribution of plasma from the blood vessels at the margin of the cornea; leucocytes also wander through the larger canals from lacuna to lacuna. Neither blood vessels nor lymphatics penetrate the substance of the cornea, nor are there transverse canals which bring into communication the

canal-systems separated by the lamellae. The surfaces of
the cornea, both anterior and posterior, are invested by
membranes in direct contact with the most superficial
lamellae ; but there is no genetic relationship between the
membranes and the lamellae beneath. Farther description
of these membranes, and of the epithelia supported by
them, will be deferred to a subsequent portion of this
work.

It should be stated before leaving the description of corneal tissue that the use of the terms lacunae and canaliculi,
as here applied to the corneal spaces and lymph channels,
is unusual. It is warranted by the resemblance between
these spaces and channels and those occurring in osseous
tissue to which these names are commonly applied; and
is offered here from the conviction that a clear idea of the
resemblances (and also of the differences) of the two tissues will aid materially in a clearer understanding of the
structure of the latter.

The terms **bone tissue** and **osseous tissue** are applied
indifferently to the chief constituent of the organs well
known as bones: like all skeletal tissues, it can be defined
by the structure of the matrix and the forms and relations
of the fixed corpuscles.

The matrix consists of thin lamellae composed in part
of fibres (the mode of whose arrangement is not clearly
demonstrable), and in part of a homogeneous ground-
substance which is strongly impregnated with lime and
other salts, calcium phosphate being the chief. Prolonged
boiling converts the fibres into a substance which has been

called **ossein**, but which is probably an impure form of
gelatin: the fibres are therefore allied to if not identical
with white fibres.

Between adjacent lamellae are found frequent lenticular
spaces of exceedingly irregular outline, the **lacunae** (formerly erroneously termed the bone-cells); like the similar
spaces in corneal tissue, these are connected together between the lamellae by branching channels, or **canaliculi**:
in bone, however, such canaliculi are not only present between the lamellae, but also penetrate them in great numbers, thus bringing into free communication lacunae of
different systems; the transverse canaliculi in some cases
traversing two or more lamellae before their termination.
The transverse canaliculi are both larger and more numerous than the interlamellar, in accordance, as will be seen,
with the mode of nutrition characteristic of osseous tissue.

Within the lacunae are found the characteristic **bone
corpuscles**: these, like the cavities which contain them,
are flattened and irregular in outline, with flattened nuclei;
they can be seen to branch in some cases, but this occurs
far less freely than in the case of the corneal corpuscles;
nor is there good evidence of the connection of the branches
through the canaliculi to any considerable extent, as has
been supposed. Leucocytes do not traverse the lymph
channels of bone, nor are any other corpuscles present.

The lamellated and calcified fibrous matrix, with its
characteristic lacunae and canaliculi, and the enclosed corpuscles, are the essential structural factors of osseous tissue wherever found: and in some of the lower vertebrates

bones occur which consist simply of a few parallel lamellae. In man and the higher animals, however, these factors are always arranged in one of two methods which
are so constant as to constitute two distinct forms of
osseous tissue called respectively **dense** and **spongy bone.**

In the former the great majority of the lamellae are arranged concentrically to narrow tubular spaces known as
Haversian canals; these contain blood vessels, lymphatics
and a small quantity of another tissue presently to be described; they are of varying length, running in a general
way parallel to the surface of the mass in which they occur, and anastomosing frequently: Each is surrounded by
several lamellae whose lacunae are put into communication with it by means of transverse canaliculi. A Haversian canal and its surrounding lamellae constitute a **Haversian system.** In the irregular spaces between the
Haversian systems of dense bone are found discontinuous
lamellae also as a rule lying in a general way parallel to
the surface of the mass: these are known as **interstitial
lamellae**; and beneath the outer surface are always found
more or fewer lamellae parallel therewith and exterior to
the Haversian systems: these are known as **circumferential lamellae.**

The circumferential lamellae in all large masses of dense
bone are pierced here and there by oblique canals for
blood vessels, etc., which communicate with the Haver-
sion canals, but differ from them in having no surrounding lamellae: these are known as **Volkmann's canals.**
They are also penetrated transversely by bundles of
white fibres (which have undergone calcification) proceed-

ing from without inwards, and by occasional bundles of
elastic fibres. The name of **Sharpey's fibres** is applied to
both, but more especially to the former. Dense bone constitutes the shafts of all long bones and forms a layer of
varying thickness on the surface of the flat and the short
bones.

In spongy bone the structure is more open, as the name
implies. The solid portion consists of a meshwork of **trabeculae**, or bars and plates of varying width and of great
irregularity of form, each of which is several lamellae in
thickness, the lamellae being in a general way parallel to
the surface and the lacunae communicating therewith by
means of the transverse canaliculi. The cavities between
the trabeculae, which are quite irregular in form and size,
are known as **Haversian spaces.** Spongy bone fills the
interior of the short and the flat bones and the ends of
the long bones.

Just as cartilage tissue is found in masses of definite
form and function (organs), which are known as cartilages, so dense and spongy osseous tissues are combined in
the definite masses which, with certain associated tissues,
make up the organs well known as bones. As in the case
of the cartilages just referred to, so in this instance it will
be expedient to describe the organs in connection with the
tissues of which they are chiefly composed.

A **bone** may be defined as a mass of osseous tissue surrounded by a proper investing and genetic membrane
known as the **periosteum** ; permeated by blood vessels and
lymphatics : and containing the tissue called **marrow.**

The **periosteum**, as found upon the surface of the bones of adult and particularly of elderly individuals, is a thin, tough, closely felted membrane which adheres tenaciously to the structure beneath; it can, nevertheless, be in most cases resolved into two distinct layers. In adolescence, childhood, and notably in foetal life these are clearly distinguishable. The outer or **fibrous layer** has the structure of an ordinary membrane, the constituent bundles (chiefly of white fibres) being closely interwoven, and the fixed corpuscles flattened, with flattened nuclei: here and there occasional fat cells may be seen, with numerous blood vessels and lymphatics, as well as delicate nerves: the outer surface frequently exhibits extensive areas having an endothelioid investment.

The inner or **osteogenetic layer** is more loosely felted, and contains a larger proportion of elastic fibres: toward the surface of the osseous tissue it contains numerous corpuscles which (particularly in young bone) are larger and more irregular in form than those of the outer layer; the protoplasm is granular and the nuclei are spheroidal: these are know as **osteoblasts,** and in youth are directly concerned in the formation of new bone, as their name and that of the layer which includes them implies. In adult bone they become flattened and inactive, forming a layer just without the osseous tissue.

The complex tissue known as **marrow** fills the internal cavities of all bones: those in the shafts of the long bones are filled with yellow marrow; the Haversian spaces of spongy bone, in most places where it occurs, with red marrow: the larger Haversian canals on the inner side of dense bone are lined with a layer of modified marrow, which is prolonged in the smaller canals by a peculiar form of connective tissue containing osteoblasts, and finally continuous on the outer surface with the osteogenetic layer of the periosteum; the marrow, as will be shown later, being originally derived from an ingrowth of that layer. The two kinds of marrow are closely related in structure, the yellow being derived from the modification of the red: the latter will therefore be described first.

The basis of structure of **red marrow** is a delicate framework of retiform tissue associated with an abundant vascular network, which will be described more fully in another connection: the interstices contain corpuscles of at least three different kinds. Throughout the mass, and particularly near the surface, are the **marrow cells,** properly so called: these are relatively large, with faintly granular protoplasm and oval nuclei; those upon the surface of the mass have the appearance and perform the function of osteoblasts in young bone. The interior of the marrow contains numerous smaller cells with granular nuclei, to which the name of **erythroblasts** has been given, on account of the part they play in the formation of colored blood corpuscles: associated with these are great numbers of the immature blood corpuscles themselves; to these the color of the marrow is largely due.

Upon the surface and more rarely in the interior of the marrow are found occasional masses of protoplasm, several times larger than the ordinary marrow cells, best known as **giant cells:** they were called **myeloplaxes** by

Robin, who believed them to be peculiar to marrow; but bodies to all appearances similar to them have since been found in other tissues. It was proposed by Kölliker to call them **osteoclasts,** from the belief that they were agents in the absorption of bone; where, as sometimes occurs in embryonic bone, the marrow lies in contact with temporary cartilage which has begun to be absorbed, it has been proposed to call them **chondroclasts** for a similar reason: this view of their function is supported by the fact that they not infrequently (but by no means invariably) lie in little pits or depressions of the surface of the bone or cartilage undergoing absorption: these pits have been called **Howship's lacunae** or **foveolae.** Nothing, however, has yet been certainly proven concerning the function of these bodies: the name first given is therefore for the present at least the most desirable.

The giant cells are of two kinds, differing chiefly as to their nuclei. Some show a large number of these bodies; others have but a single nucleus, which is always very large, and frequently of an exceedingly irregular shape: the former are termed **multinuclear** and the latter **uninuclear giant cells.** It has been supposed that the former are derived from the latter by the fragmentation of the large irregular nucleus; but Howell has shown that this view is certainly very doubtful and probably erroneous.

From the description given it will be seen that the red marrow is a very important tissue of quite complex function, having important relations not only to the surrounding bone, but also to the elaboration of the blood. The **yellow marrow,** on the other hand, is, as far as our present knowledge goes, one of the most passive tissues of the body. It differs from the red marrow, from which it is derived, chiefly by the facts that the erythroblasts and immature blood corpuscles are alike wanting, and that the great majority of marrow cells in the interior of the mass have undergone fatty transformation. Upon and near the surface both marrow cells and giant cells are found.

From the description above given it will be seen that the nutrition of dense bone is maintained by the vessels of the periosteum directly for the corpuscles of circumferential lamellae, and by the vessels of the Haversian canals for the concentric lamellae, and thus indirectly by the periosteum, from which these vessels are derived; the periosteum is thus seen to be not only a protective, but also a nutrient investing membrane: this is conspicuously shown by the fact that whenever the periosteum is removed by accident or disease from any considerable area of bone the subjacent osseous tissue perishes. In like manner the corpuscles of lamellae which make up the spicules and trabeculae of spongy bone depend upon the blood vessels of the adjacent marrow for their food supply, the Haversian spaces occupying the same relation to them as the Haversian canals to the concentric lamellae of the dense bone: these spaces are in a certain sense the expansions of the canals, as the marrow which they contain is the continuation of their lining, and thus, in a roundabout way, of the inner layer of the periosteum. It is also to be noted that the lamellae in each case depend for their nutrition upon supplies drawn from a surface to which they are parallel: the importance of the transverse canaliculi and the reason for their number and extent thus becomes evident.

In speaking of tendon tissue it was stated that while a mechanism of interstitial increase was known to exist, we had no clear evidence of any continuous process of interstitial removal and replacement. The case of bone is different: no one can examine a transverse section of dense bone, and note the manner in which what appear to be recent Haversian systems cut into the territory of what are probably older, and how both the interstitial and the circumferential lamellae are interrupted by both, without the conviction that a process of replacement is involved: and this conviction is confirmed when we compare the cross section of the femur (for example) of a child with that of an adult, into the central marrow space of which it could be thrust, each with its Haversian systems, its interstitial and its circumferential lamellae. Dense bone shows us, however, no tissue in process of removal, and no Haversian systems in process of formation: and the means by which both removal and replacement take place are yet to be discovered.

CHAPTER VI.

OSSIFICATION.

Attention has already been called to the fact that the periosteum is at once an investing and nutrient, and a genetic membrane; and to the terms applied to its inner layer and the corpuscles contained therein. It has also been pointed out that the marrow cells upon and near the surface of the red marrow agree with osteoblasts in form and also in function. The function of an osteoblast may now be briefly stated by saying that it is its normal destiny to become a bone corpuscle. Bone is laid down in lamellae by the activity, probably periodical in character, of an osteogenetic layer, the osteoblasts nearest the lamella just previously formed being enclosed by the new layer of fibrillated and calcified matrix, and thus converted into bone corpuscles.

It has already been stated and should always be born in mind that the marrow, as will shortly be seen, is a development from the osteogenetic layer of the periosteum: its outer surface, with its layer of osteoblasts, is as truly an osteogenetic layer as that from which it is derived: attempts have, indeed, been made to define it as a distinct membrane under the name of the **endosteum**; but this is unnecessary: and is not advisable, if for no other reason,

on account of the fact that it cannot be anatomically separated from the tissue below, nor its boundary on that
side at all clearly defined.

The formation of osseous tissue by the enclosure of osteoblasts in a fibrillated and calcified matrix is called **ossification**, wherever it occurs, as, for instance, in the bones
of a growing child or young animal. The term is more
especially applied, however, to the original formation and
early development of bone as it takes place in the embryo.
Used in this sense, the name is applied to two processes
often regarded as essentially different, and designated respectively as **intramembranous ossification**, or the development of bone in previously existing connective tissue,
and **ossification in cartilage**, or the development of bone
in previously existing cartilage. It is of the highest importance, however, to keep clearly in mind the fact just
stated, that bone is always formed in a peculiar kind of
connective tissue, the osteogenetic layer already defined as
in every instance derived directly or indirectly from a membrane. Bone is therefore always derived from the modification of a membrane: and the facts of comparative anatomy and of pathology, on the other hand, show that
almost any membrane may under certain circumstances
become osteogenetic.

The so-called ossification in cartilage, or, as it is often
termed (and the expression is still more apt to mislead),
the ossification of cartilage, is in reality the **replacement
of cartilage** by bone, the cartilage itself being absorbed
and disappearing in great measure before bone formation takes place. It is particularly important to avoid

the erroneous impression that is sometimes caused by the expression in question, to the effect that the previously existing cartilage is in some manner transformed into bone: if this be borne constantly in mind, there is no objection to the use of the term in either form; nor, with this qualification, to the use of the terms **"membrane bone"** and **"cartilage bone,"** frequently employed for convenience to designate briefly bones formed respectively in the two methods above defined.

Since bone is to be regarded as the result of a development originating in connection with a membrane, we may properly first consider the process of **intramembranous ossification**: this takes place in the human body and in that of the higher vertebrates generally in a very small number of bones, the great majority of those entering into the structure of the adult skeleton being preceded by cartilage: the tegmental bones of the skull, the squamous portion of the temporal bone, the bones of the face and jaws (excepting a small portion of the lower jaw), and the clavicle are the only bones not so preceded, or, in other words, the only so called membrane bones.

A study of the development in the embryo of any one of these bones gives substantially the following results. The place of the future bone is at first occupied by a mass of embryonic connective tissue not yet possessing the density and structure of membrane, permeated by a network of blood-vessels. At one or more places, known as **centres of ossification**, bundles of stout fibres are formed, radiating outward from a point midway between the adjacent vessels: these, which are known as **osteogenetic fibres,** resemble white fibres in appearance, but are less distinctly fibrillated: between them, and crowded upon the surfaces of the bundles, are large numbers of connective corpuscles, modified to form osteoblasts. The homogeneous substance which forms their matrix now begins to undergo calcification, the salts being at first deposited in the form of minute globules; as these become more and more numerous, they fuse together forming a continuous and apparently homogenous mass: the osteoblasts imbedded therein becoming the bone corpuscles, and the modified bundles of fibres becoming spicules of bone.

The formation of new fibres, with their associated corpuscles, continues to go on in advance of calcification at the extremities of the spicules, the tufts of bundles diverging in such manner as to anastomose with those of adjacent spicules, their growth taking such direction as to pass between the meshes of the vascular network. As a result, there is formed a spongy network of bone interlacing with the network of blood vessels, the interstitial embryonic connective tissue forming the basis of the primary marrow which fills the spaces and in which the blood vessels are imbedded. While this is taking place in the interior of the mass, the embryonic connective tissue upon its surface becomes converted into a well defined layer of fibrous membrane, the fibrous layer of the periosteum: from the osteogenetic tissue immediately beneath dense periosteal bone is formed. The ossification thus set up is continued outward from the centre or centres of ossification until the whole territory involved is converted into osseous tissue, with the associated periosteum, marrow and blood vessels: as the bone increases in thickness new layers of dense bone are deposited by the periosteum, while that first formed is absorbed and replaced by the spongy bone of the interior.

In the replacement of cartilage by bone, or, as it has been sometimes termed, **intracartilaginous ossification,** blood vessels also play a conspicuous part, as we shall presently see; the centres of ossification in this, as in the preceding case, arising by the development of highly vascular areas of osteogenetic tissue: the formation of these areas and their subsequent extension, known as the **vascularization of cartilage,** is the immediate preliminary to the deposition of bone. The process is rendered more complex by the fact that the previously existing cartilage must necessarily be removed, at least in great measure, before bone can be deposited in its stead; and the additional fact that the receding cartilage is itself the seat of noteworthy changes which invariably take place, although their direct relation to the formation of bone is by no means clear. It should also be mentioned, before beginning a detailed discussion of the process in question, that the replacement proper of the cartilage is also accompanied or followed at a very early stage by its investment with bone formed by the newly developed periosteum; and that this takes place at the outset in a manner somewhat resembling the process of intramembranous ossification just described; whereas, at a later stage, the formation of dense bone takes place.

We may, therefore, recognize four distinct and definite stages in the process of intracartilaginous ossification, so-called (to the second of which alone, however, the term is strictly applicable): these stages are in the main successive for any particular point; though all may be in progress at the same time at points adjacent to each other; they are definable as follows.

The first, which may be called the **transformation of cartilage,** includes all those changes which take place in that tissue from the first disturbance of its normal condition to its final dissolution in great measure. The second is the development of spongy bone in the spaces formed by the dissolution of the cartilage and upon its remains, or the **formation of endochondral bone.** The deposition of spongy bone beneath the newly developed periosteum as an investment of the endochondral bone and the transforming cartilage is the third: from the position where it occurs this is designated the **formation of perichondral bone,** or, as it is sometimes called, primary periosteal bone. The fourth and last stage is the **formation of dense bone** surrounding the perichondral bone, through the continued activity of the periosteum; the transition from the one to the other being in some cases exceedingly gradual.

We may now enter upon a discussion of the changes which take place in the formation of the shaft of one of the long bones, such as the humerus, the tibia, or one of the metatarsal bones. The bone is preformed, to use a current expression, in cartilage: that is to say, its future place is occupied by a mass of hyaline cartilage having in a general way the form and relations of the bone that is

to replace it: the mass of cartilage is covered by a single
and closely adhering fibrous layer, the perichondrium,
which is moderately rich in blood vessels. The transformation of the cartilage involves four recognizable changes,
which may be designated as the **rearrangement of the
corpuscles**, the **calcification of the matrix**, the **formation of primary areolae**, accompanied by the degeneration of the corpuscles, and the **formation of secondary
areolae** by the partial dissolution of the matrix.

At a point near the centre of the mass the corpuscles begin to multiply and to increase in size, and (in a manner
not yet clearly understood) to arrange themselves in
columns or rows which at first radiate from the central
point, with intervening regions consisting of matrix only,
thus forming a spheroidal region of transformation which
continues to increase: when the surface of the mass is
reached on the adjacent sides the region of transformation
is of course restricted to the cartilage lying in the direction of the ends: and as it advances toward them in both
directions the columns of rearranged corpuscles and the
intervening regions of matrix soon take on a direction
parallel to the axis of the bone.

Very shortly after the rearrangement of the corpuscles
just described, the deposition of lime salts in the matrix
takes place, particularly in those tracts which lie between
the corpuscular rows, bars and plates of calcified cartilage
thus being formed. At the same time the cavities in which
the cartilage corpuscles lie begin to be enlarged, while the
corpuscles themselves undergo degeneration, becoming
shrunken and irregular in form, and lying in the enlarged

cavities, which are now known as the **primary areolae**
above referred to. In some cases the areolae consist of two
or three cavities in the same row, united by the dissolution
of the thin lamella of matrix lying between them. The
farther dissolution of the matrix, and the formation of the
secondary areolae, in which the deposition of endochondral bone begins, depends upon the development of other
structures presently to be described.

About the time that the transformation of cartilage is
set up in the centre of the mass important changes take
place in the surrounding membrane, beginning in a zone
immediately adjacent, and proceeding thence toward either
extremity simultaneously with the advancing transformation of the mass within. The fibrous layer becomes
thicker and more highly vascular: its inner portion becomes looser, and the corpuscles rapidly increase in number and in size, with associated changes in form and
structure: in other words, the perichondrium of the embryonic cartilage becomes converted into the periosteum
of the future bone, with an outer fibrous and inner osteogenetic layer.

By the time the central spheroidal area of transforming cartilage has reached the lateral surfaces of the mass,
the adjacent zone of periosteum is fully formed; and the
deposition of a delicate network of spongy bone may have
already taken place. As the two regions of activity come
in contact, one or more loops of blood vessels grow out
toward the transformed cartilage, carrying with them an
investment of osteogenetic tissue: before their advance the
transformed cartilage is absorbed, and they quickly reach

the centre of the mass. This is sometimes termed the **primary vascular invasion.** From the centre the formation of vascular loops and the associated development of the investing osteogenetic tissue is directed toward the ends of the bone, following the advancing areas of transformation of cartilage; the thin lamellae of matrix lying between the primary areolae in the longitudinal rows are absorbed, as well as (to some extent) those lying between adjacent rows: there is thus formed a network of elongated spaces, the **secondary areolae,** separated by intercommunicating bars and plates (trabeculae) of calcified cartilage matrix, and opening toward the centre of the bone.

These spaces are rapidly filled by the advancing blood vessels and associated osteogenetic tissue, which together form the primary marrow: the surface of the mass is invested with numerous osteoblasts, which are brought directly in contact with the trabeculae of calcified cartilage matrix: upon the surfaces of the latter the deposition of thin lamellae of true bone substance begins. There is thus formed a spongy mass of bone trabeculae having a small quantity of calcified cartilage included in their structure, and forming an area which advances towards the end of the body just behind the area in which the last steps in the transformation of the cartilage are taking place: this is the **endochondral bone.** In the long bones with hollow shafts it is itself a more or less temporary structure, being absorbed from behind almost as rapidly as it is formed along the area of advance, its place being taken by the **permanent marrow.** This is formed from the modifi-

cation of the primary marrow, which at first consists
only of the blood vessels and the associated osteogenetic tissue, which is quite delicate and almost gelatinous
in structure: later, its retiform framework becomes more
fully developed, giant cells and erythroblasts appear, and
the characters of ordinary red marrow are assumed; these
in turn (in the region referred to) giving place to those of
the yellow marrow which finally occupies the cavity.

It has already been stated that the newly developed zone
of periosteum begins shortly after its formation to deposit a layer of spongy bone upon the surface of the transforming cartilage in the middle of the shaft, similar to that
first formed where the seat of ossification is a membrane.
Along with the progress of the areas of the transforming
cartilage toward the extremities of the bone, and the immediately successive development of endochondral bone in
the interior of the mass, there is a corresponding advance
in both directions of the change which converts the fibrous
membrane investing the cartilage into periosteum. This,
again, is immediately followed by the deposition of a layer
of spongy bone, or, rather, by the extension of the layer
already begun. While the layer so formed is in this manner steadily increasing in extent, it also undergoes increase
in thickness in the region where it was first formed.
There is thus formed a sheath of spongy bone, shaped like
a dice box open at both ends, thickest in the middle and
gradually becoming thinner towards its extremities, which
contains in its middle the newly formed permanent marrow: its ends surround the masses of endochondral bone
previously described; immediately beyond lies the area of

transforming cartilage: and still beyond (and invested as yet by perichondrium) the hyaline cartilage which has thus far undergone no change. This sheath of spongy bone is the **perichondral bone,** or as it is sometimes called, the primary periosteal bone.

As the periosteum extends its borders toward the ends of the future bone, its newly formed osteogenetic layer making the deposit of perichondral bone just described, a change, more or less gradual, takes place in the activity of its central and older portion. The meshes of the vascular network within its inner layer become more and more elongated in the direction of the long axis of the bone: the osteogenetic tissue which surrounds the vessels, instead of continuing to occupy a large portion of the intervascular spaces as marrow, deposits successive lamellae of bone substance concentric to the vessels, until the spaces are reduced to slender canals. Haversian systems are thus developed; while from the immediate surface of the periosteum interstitial and circumferential lamellae are produced, and the sheath of perichondral bone is itself invested with a layer of **dense bone,** or permanent periosteal bone. The perichondral bone is but little more enduring than the endochondral bone which it for a time surrounds: it has already been stated that as the latter advances toward the degenerating cartilage, it is itself, in the long bones, sooner or later absorbed and replaced by the permanent marrow: the perichondral bone suffers a like fate, the mass of marrow becoming enlarged in diameter as well as in length, and the surrounding spongy bone being absorbed about it until the marrow lies directly beneath the tube of dense bone which forms the permanent shaft of the femur or tibia, as the case may be.

It will be remembered that the mass of cartilage which in the foetus represents one of the long bones under consideration, is very much smaller than the bone which we find in its place even in the newborn child: the increase in thickness will be readily understood. It should be borne in mind, however, that this increase in thickness is for quite a while due to the formation of bone of the perichondral type only: it is not until some months after birth that dense bone of the ordinary type is deposited, the shafts of the long bones remaining up to that time more or less spongy throughout.

The increase in length is due to a continuous growth of the primary cartilaginous mass itself, which up to birth and for some time after continues to increase in length and also in thickness in the part which has not as yet begun to undergo transformation. This growth continues at a constantly decreasing rate, the region of untransformed cartilage thus becoming smaller and smaller. It is encroached upon alike by the advancing endochondral bone of the shaft, and the similarly formed but more permanent mass of spongy bone in the epiphysis, until finally the two areas of ossification meet, and the union of the epiphysis with the shaft is accomplished, the two masses of spongy bone and their investing layers of dense bone alike becoming confluent.

The table on the opposite page presents a summary of the various processes which have been described above as taking place in the formation of the shaft of a long bone.

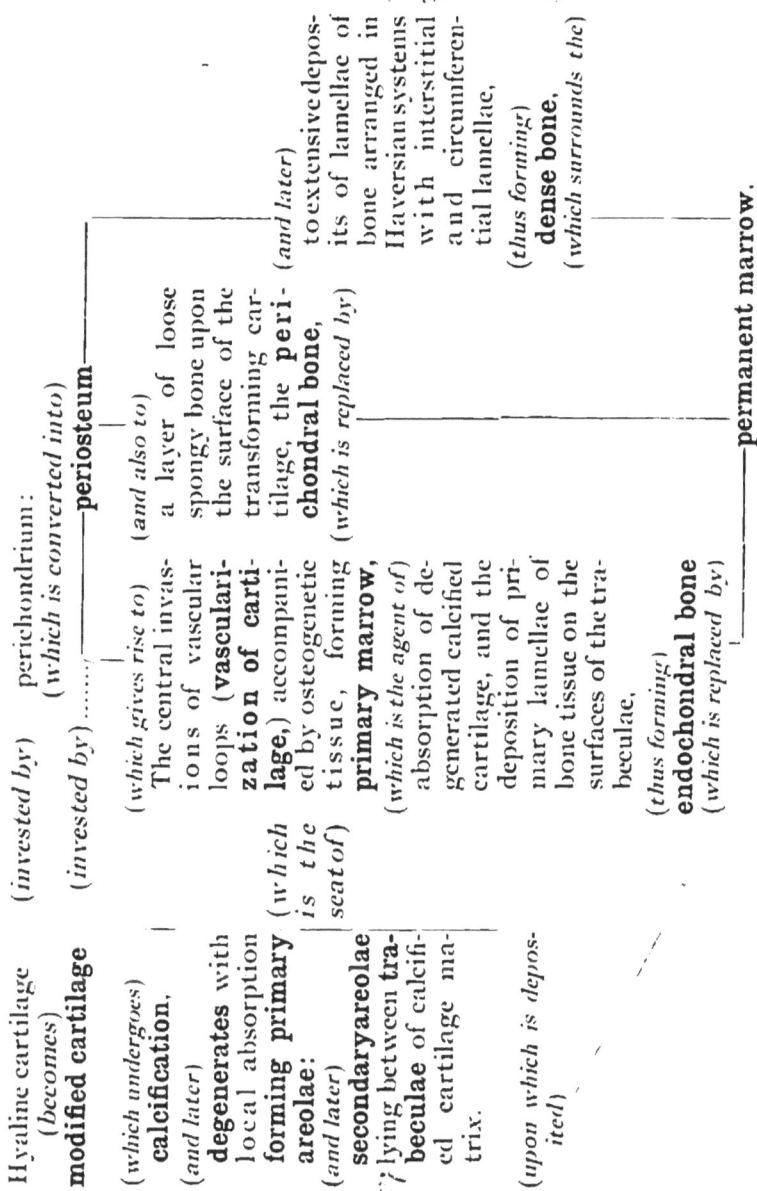

Hyaline cartilage
(becomes)
modified cartilage

(invested by) perichondrium: *(which is converted into)*
(invested by) .... **periosteum**

(which undergoes) **calcification,** *(and later)* **degenerates** with local absorption **forming primary areolae:** *(and later)* **secondary areolae** "?lying between trabeculae of calcified cartilage matrix.

(upon which is deposited)

(which is the seat of)

(which gives rise to) The central invasions of vascular loops **(vascularization of cartilage,)** accompanied by osteogenetic tissue, forming **primary marrow,** *(which is the agent of)* absorption of degenerated calcified cartilage, and the deposition of primary lamellae of bone tissue on the surfaces of the trabeculae,

(thus forming) **endochondral bone** *(which is replaced by)*

(and also to) a layer of loose spongy bone upon the surface of the transforming cartilage, the **perichondral bone,** *(which is replaced by)*

(and later) to extensive deposits of lamellae of bone arranged in Haversian systems with interstitial and circumferential lamellae, *(thus forming)* **dense bone,** *(which surrounds the)*

permanent marrow.

In the short bones, and in the epiphyses of the long bones, the replacement of cartilage takes place in an essentially similar manner, the vascular invasions reaching the centre of the mass of transforming cartilage, and the development of spongy endochondral bone taking place on all sides. Some of the trabeculae thus formed are shortly afterwards absorbed, while others become larger, and form part of the permanent framework of the interior of the bone. It is probable even here, however, that there is in the earlier part of adult life at least more or less of absorption and rebuilding continuously going on. An investing layer of perichondral bone is formed, and contributes to the permanent spongy mass: and later a thin layer of dense bone is superimposed. There are some very interesting variations in detail as to the manner in which ossification arises in some of the bones, but their consideration would be foreign to the purpose of this work.

CHAPTER VII.

THE BLOOD.

The **blood** consists of a fluid portion, the **plasma**, and numerous **corpuscles** found floating therein. It has sometimes been described as a tissue with a fluid matrix. It is perhaps a straining of the use of these terms to apply them to the blood: it is certain, however, that a description of the structural elements of the body would be incomplete without an account of those contained in that fluid. While they cannot be strictly defined as connective tissue elements, their origin is such as to warrant their description at this time.

The **plasma** is during life and in health a perfect fluid as long as it is contained in the blood vessels. When allowed to escape therefrom, or more rarely under certain abnormal conditions while contained therein, it undergoes coagulation, an important process, which concerns us chiefly because of the formation of a solid substance whose appearance should be familiar to the student of histology. It is known as **fibrin,** and when present in small quantities can be plainly seen to consist of exceedingly delicate filaments interlacing in every direction; as the quantity increases, the meshes of the network thus formed become filled with fibres, until a practically continuous solid mass

is formed: when hardened and stained such a mass presents in section a granular appearance due to the cross-
section of the fibres.

The corpuscular elements contained in the plasma and
constituting normally a little over one-third of the volume of the blood, are chiefly of two kinds, the **colored** (or
so-called **red**) **corpuscles,** or, as they have been termed,
the **erythrocytes,** and the **colorless** (or so-called **white**)
corpuscles, which are also commonly known as **leucocytes.** There are also present in normal blood smaller or
larger quantities of minute bodies known by various
names, the most common of which is the **blood platelets.**

The **colored corpuscles** are by far the most numerous,
about five millions being contained in a cubic millimetre of
the blood of a healthy man: the number in the blood of
women is about ten per cent. less: in certain forms of disease their number may be very greatly reduced. When
seen singly under the microscope each colored corpuscle is
seen to be of a yellowish green color, the former tint pre-
vailing in arterial, the latter in venous blood. The form
of each is that of a biconcave circular disk, the central
portion being slightly hollowed on each side and the margin rounded.

The size of the colored corpuscles varies within rather
wide limits, the same sample of human blood exhibiting
individual corpuscles whose transverse diameter is as
great as ten micra, and others that are but half as broad:
by far the great majority, however, will be found to be

from seven to eight micra in breadth: the average measurement has been variously stated by different observers after a very large number of measurements; it is not far from the truth to say that it is about seven and three-fourths micra. Little difference, if any, is found in this respect in the blood of persons of different age or sex, nor is any noticeable in that of the various races of mankind. The greatest thickness of the blood corpuscle is about one-fourth its breadth. The smallest corpuscles found are sometimes distinguished as a separate form of elements under the name of **microcytes**; but this distinction is of doubtful significance.

The colored corpuscles of the blood of all mammals save those of the camel family resemble those of human blood in form and general appearance: they vary, however, greatly in size in different species, the differences that are found to occur having no relation to the differences in size of the animals themselves; those of the mouse, for example, being distinctly larger than those of the horse. The largest known are those of the elephant (between nine and ten micra in diameter) and the smallest those of the musk deer (about two and one-half micra). There is, also, no necessary close resemblance in this respect between nearly allied animals; thus, while the colored corpuscles of the blood of oxen, sheep and goats are very nearly of the same size, those of dogs and cats differ greatly.

While the colored corpuscles of the blood of many mammals differ in size so decidedly from those of human blood as to make their distinction a matter of absolute certainty, there are several species of mammals whose colored blood

corpuscles approximate so closely in size to those of man
as to render impossible a certain distinction between them.
While, therefore, it might be possible under certain circumstances to testify that a given stain was not caused by
human blood, the converse is not justifiable in the present
state of our knowledge.

The structure of the colored corpuscles has been and still
is a matter of much debate. By some the disk is regarded
as consisting of a denser colorless portion, spongy in structure, the **stroma**, whose meshes contain the fluid colored
portion, a solution of the substance known as **haemoglobin**: its behavior towards certain reagents (notably toward water) leads others to the conviction that what we
have is really a closed sac (or cell in the true sense of the
word) filled with the colored fluid. It is at least certain
that the corpuscle is far from being a mass of protoplasm,
being as greatly modified therefrom as a fat cell. It should
also be noted that a nucleus is wanting in the ordinary
colored blood corpuscles of all mammalia: in which respect they differ from those of all other vertebrates.

When blood is drawn from the vessels, or when, from
any cause, it stagnates for a time within them, the colored
corpuscles show a tendency to adhere together by their
sides in masses resembling piles of coin; this **formation of
rouleaux**, as it is termed, was once supposed to be peculiar to blood corpuscles; but it has been shown that under
certain circumstances the same process may be caused to
take place with floating disks of cork or other substances
under conditions that are purely mechanical. The rouleaux thus formed not unfrequently arrange themselves in

a coarse network in whose meshes the colorless corpuscles may be seen, and the first formed filaments of fibrin
detected as coagulation begins. The formation of rouleaux does not take place in the blood of vertebrates other
than mammals, the thickening of the centre of the corpuscles due to the presence of the nucleus acting as an
obstacle thereto.

If sodium chloride or any other salt, sugar, glycerine, or
any other reagent that tends to increase the specific grav-
ity of the plasma be added to fresh blood, or if the same
effect is produced by circumstances favoring rapid evapo-
ration, the colored corpuscles undergo a characteristic
change of form known as **crenation**, the corpuscle becoming shrivelled and the surface studded with minute projections, the whole being in appearance not unlike the fruit of
a thorn-apple or a horse-chestnut. This change may possibly take place in the vessels themselves under circumstances temporarily causing an appreciable difference in
the proportion of water in the blood.

The **colorless corpuscles**, or **leucocytes**, are nucleated
masses of protoplasm which may be regarded as typical
animal cells. When at rest they are spheroidal in form,
but, as will be seen later, are capable of very great modification in this respect. They are far less numerous than
the colored, a cubic millimetre of healthy human blood
containing on an average about ten thousand : the number
varies greatly, however, not only in different individuals,
but in the same individual under different conditions;
thus, the number is greatly increased shortly after eating

and markedly diminished by prolonged fasting: we may therefore find in the blood of a healthy person at one time twice as many to the cubic millimetre, at another even less than half as many as the average above given. Since the number of the colored corpuscles is not subject to such marked fluctuation, the ratio between the two varies therewith; and the number of colorless corpuscles is often stated in terms of that ratio as ranging between one colorless to two hundred and fifty colored and one colorless to over one thousand colored corpuscles: the average being about one colorless to five hundred colored corpuscles. This mode of statement is, however, unsatisfactory, for the reason that, as will be readily seen, a like change of ratio would result from a marked diminution in the number of colored corpuscles or an undue increase in the number of the colorless, changes in each case of very great importance, but of quite different significance.

The colorless corpuscles vary also in size to a great extent: their average diameter, when in the spheroidal state, may be stated as about ten micra. The smallest, sometimes distinguished specifically as **small leucocytes**, have large spheroidal nuclei, surrounded by a small amount of protoplasm: they may be regarded as newly-formed elements that have only recently entered the blood stream; they are somewhat smaller than the average colored corpuscles. When fully developed they may become as much as twice as large, the increase being chiefly in the amount of protoplasm: they are then sometimes termed **large leucocytes**; still retaining the single spherical nucleus.

The smaller and larger **uninuclear** elements make up in all about twenty-five per cent. of the leucocytes of the blood. Almost all of the remainder (about seventy per cent. of the whole number) are what are sometimes termed **multinuclear leucocytes**: these, which are comparatively uniform in size and somewhat larger than the colored corpuscles, have either two or three small nuclei or one large nucleus of irregular form and apparently about to undergo fragmentation. In addition there may be found (occurring but sparsely in normal blood) a small number of leucocytes with pale nuclei and highly granular protoplasm, the contained granules staining deeply with eosin and similar reagents. These are sometimes known as **granular** or (from their affinity for the substance just mentioned) **eosinophilous leucocytes.**

When human blood or that of any mammal is examined at an ordinary temperature the colorless corpuscles are spheroidal and motionless: if, however, the stage of the microscope be so heated as to maintain the temperature of the blood at that of the living body, the spheroidal form is no longer maintained. The living corpuscle puts out one or more stout lobular processes, thus assuming an irregular form bounded by sweeping curves: these curves are seen to change slowly but constantly, the outline not remaining the same for any length of time. This shifting of form may go on about equally in all directions, the corpuscles as a whole remaining stationary; or there may be a flowing movement of the protoplasmic body toward one of the lobular processes, and a consequent change of position, as irregular and indefinite as the movements by

which it is caused. This mode of motion, characteristic of leucocytes wherever found (and of some other
elements as well) is identical in its nature with that seen
to take place in the members of the genus **amoeba,** a
group of very lowly organized unicellular animals: it is,
therefore, termed **amoeboid.**

All the other forms of colorless corpuscles are probably
derived from the modification of the small unicellular elements. These, in turn, are brought into the blood stream by
the lymphatics, where, under the name of **lymphocytes,**
they constitute the corpuscular elements of the lymph,
some, however, attaining their full size while in that fluid.

The origin of the lymphocytes is well known. They are
formed chiefly, if not solely, in those organs which consist
in great measure of what is known as **lymphoid** or **adenoid tissue:** those known as **lymphatic glands,** or, more
properly, as **lymphatic nodes,** having the formation of
lymphocytes as their principal if not their sole function.
The histological anatomy of these organs will be described
in a subsequent chapter: it is sufficient here to say that
they consist in the main of masses of adenoid tissue well
supplied with blood vessels and enclosed in each instance
in a capsule, into which several small lymphatic vessels
enter and from which a single larger lymphatic vessel
leads, the current of lymph which thus passes through the
mass, carrying with it the newly formed lymph corpuscles.

What is here termed **adenoid tissue** is, like the marrow
of bone, a compound structure of which retiform tissue is
the basis. In addition to the flattened connective tissue
corpuscles which form an endothelioid layer upon the reti-

culum, and which have been described in a previous chapter as the fixed corpuscles proper of retiform tissue, the
trabeculae which make up the framework of the mass are
⟡ in most cases so numerous as to entirely fill the meshes of
the network. These cells, properly known as **lymphoblasts,** multiply rapidly by cell division, the older gradually falling into the lymph channels which penetrate the
mass and becoming young lymph corpuscles or lymphocytes. At this stage they are spheroidal, with large nuclei
and a very small amount of investing protoplasm: the
latter substance increases in quantity as they are carried
to the heart; but, as we have already seen, many enter
the blood stream before they are fully matured.

The origin of the colored corpuscles of the blood is now
also well established. During the early stages of embryonic life large nucleated colored corpuscles are formed in
the newly forming blood vessels. Later, colored corpuscles of the ordinary type are formed first in the liver, later
in the spleen, and finally in the red marrow. After birth
the last named structure is the chief and probably the sole
place of their formation in most if not all mammals, at
least during health; it is possible that their formation
may be temporarily resumed by the spleen if not by the
liver under certain abnormal conditions.

In the description of the marrow already given it was
stated that the interior of that structure contains numerous small cells with granular nuclei known as **erythroblasts**: these are formed by cell division from larger cells
not distinguishable from ordinary marrow cells. The
erythroblasts so formed multiply rapidly by the ordinary

method of cell division, the cells thus formed being spheroidal and nucleated and gradually becoming converted
into immature blood corpuscles by the formation within
them of haemoglobin. Later the nucleus is extruded from
the corpuscle and the spheroidal mass becomes converted
into the biconcave disk which we find in the blood stream.

Reference has been made to the **blood platelets.** These
are minute rounded and colorless bodies (from one-third
to one-fourth the diameter of a colored corpuscle) which
are found in the blood either singly or adhering together
in masses of considerable size: their total number in normal human blood is but small. They have been described
under a number of names, as the **elementary particles**
of Zimmermann, the **granules** of Osler, the **haemato-
blasts** of Hayem. The name of blood platelets was proposed for them by Bizzozero. They have been regarded
by Hayem and others as concerned in the formation of red
corpuscles within the blood stream, but the evidence for
this view is not conclusive. Perhaps the most probable
explanation of them is that they are the disintegrating
fragments of broken down colorless corpuscles. Our
knowledge of them is, however, exceedingly imperfect.

CHAPTER VIII.

THE CONTRACTILE TISSUES.

———

What is in physiological language termed contractility consists in a change of form but not of volume. Irregular contractility, or change of form in any direction and in an indefinite manner, is one of the powers which are inherent in living matter and may be manifested by any cell which is still in an embryonic condition: that specialization of function and structure which converts the embryonic cell into the tissue element is in many cases accompanied by the disappearance of this power: it is retained, however, in some cases, as in the contractility of pigment corpuscles in many vertebrates, and notably in the amoeboid movements of leucocytes which has been described in the preceding chapter.

In one group of elements, however, this power is specialized and becomes their distinctive function. This specialization is of more than one kind: there is probably an increase in the actual amount of contraction, and certainly an increase in its rate; but the most important feature is its definiteness of direction, one axis of the mass (the longest) always becoming shorter, while the mass as a whole becomes thicker. From the fact that elements belonging to this group are the essentials of structure of the organs known as muscles, the tissues formed of them are known

as **muscular tissues,** although the elements of which they
are composed sometimes occur singly and are frequently
found in masses that cannot with propriety be called muscles.

There are three distinct kinds of muscular elements or
fibres (as they are commonly termed, from the elongation
generally characteristic of them) the smooth, the cardiac,
and the striped or striated muscular fibres. The elements
of the first two kinds have in each instance a single nucleus, and may therefore be regarded as resulting from the
direct modification of a single embryonic cell. Those of
the third kind are much longer as well as thicker than the
others, and while they arise in each instance from a single
embryonic cell, this becomes greatly elongated and the
nucleus divides repeatedly; the resulting strand of protoplasm thus becoming multinuclear.

What is variously called **smooth, unstriped, plain** or
involuntary muscular tissue is composed of spindle
shaped cells or fibres whose protoplasmic bodies show at
times quite distinct evidences of longitudinal striation,
but are at other times perfectly plain. The existence of a
very delicate investing membrane or sheath has been demonstrated. The nucleus is elongated, sometimes oval but
in many cases distinctly rod shaped, and is situated in the
centre of the mass. Smooth muscular fibres vary somewhat in size and particularly in length: the transverse
diameter usually ranging between five and ten micra,
while the length may be less than ten or more than twenty
times the diameter. When so situated as to escape lateral

pressure the smooth fibres are circular in cross section: when, however, as is frequently the case, they are pressed together, the sides become flattened and their cross sections polygonal. When seen in transverse section, therefore, the investing membrane forms a circle or polygon, within which is seen the protoplasm of the body of the fibre, devoid of any clearly discernible structure, and in the centre the circular section of the nucleus.

Smooth muscular fibres are usually associated in bundles, the tapering extremities (which are sometimes forked) overlapping upon the bodies of adjacent fibres and adhering closely thereto; a small amount of intercellular cement substance intervenes, as can be demonstrated by the use of silver nitrate. Little if any skeletal tissue pervades the bundles. The latter are in some cases more or less loosely interwoven: their most common arrangement, however, is in more or less extensive layers; as, for instance, in the muscular wall of the intestine: such layers are penetrated by areolar tissue accompanying the blood vessels and nervous supply of the muscle fibres. It is very rarely the case that smooth muscular fibres are aggregated together into definite masses that can with propriety be called muscles, their most common occurrence being in the blood vessels and the viscera. In no case are they under the control of the will.

The **cardiac fibres** are found, as their name implies, in the muscular substance of the heart, both the auricles and the ventricles being chiefly composed of them : they also constitute an important portion of the walls of the pulmonary veins and the superior and inferior venae cavae for a short distance previous to their openings into the auricles. What are known anatomically as fibres are in this case as in others aggregates of structures not visible to the naked eye. In this instance, however, a confusion sometimes arises from the application of the term fibre to bodies which are thus compared with ordinary striped fibres: they are in reality rows of shorter elements more nearly comparable to the smooth muscular fibres.

The **cardiac muscular elements** are short, stout, irregularly prismatic bodies, intermediate in diameter between smooth and striped fibres and three or four times as long as wide. Their ends in some cases terminate squarely, in others are quite jagged and irregular. They are sometimes of uniform diameter throughout, but many give off short branches which unite with those from adjacent elements. Nothing like an investing membrane or sarcolemma has been demonstrated. The elements are faintly striated longitudinally and more distinctly transversely. Each element has a single nucleus, surrrounded by a comparatively large amount of protoplasm showing no trace of the structure which in the superficial portion gives rise to the appearance of striation.

The cardiac elements are joined together by their ends to form the cardiac fibres to which reference has been made, a larger or smaller quantity of intervening cement substance being clearly demonstrable. As many of the elements branch and anastomose with those of adjacent fibres, the appearance presented is that of a network with elongated and narrow meshes.

The various names of **striped, striated, voluntary** or **skeletal muscular fibres** are applied to those multinucleated fibres which form the organs ordinarily called muscles and usually attached to the bony or cartilaginous skeleton, the majority of them being under the control of the will. They are much larger than the smooth or the cardiac muscular elements, their transverse diameter ranging from ten to seventy micra, while they are in some cases as much as three or four centimetres long: they are prismatic in form, the ends tapering more or less gradually. Each fibre is invested by a thin homogeneous membrane known as the **sarcolemma**: within this is the mass of modified protoplasm which is the seat of the function of contraction: its most conspicuous feature is the transverse striping or striation which gives to the fibres the name most commonly applied. This, when seen by moderate powers of the microscope, presents to the eye the appearance of alternating dim and clear bands; while through the middle of the clear band may be seen a narrow black line: a longitudinal striation may also be seen, but usually less distinctly.

Beneath the sarcolemma, lying between it and the contractile substance, may be seen here and there elongated oval nuclei: these may be shown to be surrounded with a small amount of granular protoplasm which extends as a thin disk for a short distance around the nucleus. The protoplasm and the nucleus together make up what is known as a **muscle corpuscle**: these, like the similar masses in the centre of the cardiac elements, may be regarded as the residuum after the formation of the contractile substance. When a cross section of a bundle of striped muscular fibres is examined the nuclei are seen between the sarcolemma and the contractile substance, the latter being subdivided into small polygonal areas, the **fields** or **areas of Cohnheim.**

The appearances above described are easily seen: the explanation of the structure of the contractile substance upon which most of them depend is still a matter of dispute. The following facts are, however, quite generally conceded, and will probably form the basis of any further positive addition to our knowledge. The contractile substance may be regarded as made up of a clear viscid or semi-fluid portion, the **sarcoplasm**: imbedded in this are great numbers of elongated or rod-like bodies (whose exact form is not yet certainly demonstrated, and probably varies with different animals); these are known as the **sarcous elements.** They are quite uniform in length and lie in disk-like groups which compose the dim bands or zones of the fibre, the clear zones being filled chiefly by sarcoplasm : the exact cause of the dark line in the middle of the clear zone (known as **Dobie's line** or as the **membrane of Krause**) is not yet certainly known. The sarcous elements are not only regularly grouped across the fibre, but also succeed each other regularly along its length, and are possibly united end to end; the rows of sarcous elements constituting the **fibrillae.**

The sarcous elements are not uniformly distributed across the fibre, the fibrillae which they compose being grouped together in strands known as **muscle columns:** These are separated from each other by sarcoplasm, and the columns themselves are somewhat irregularly aggregated in a similar manner. This may best be seen in the cross section of a fibre, where the areas of Cohnheim are the cross sections of the columns, the finely granular appearance of their interiors representing the ends of the fibrillae; the lines which bound the areas are composed of sarcoplasm, those which are thickest separating the groups of columns above mentioned.

The conversion of the whole of the interior of the fibre into sarcoplasm and sarcous elements and the consequent lateral position of the muscle corpuscles is characteristic of most striped fibres of adult mammals, if not of all. In the young of most mammals, however, and particularly in the embryo, this conversion is not entire, and the nuclei are still found in the interior of the fibre: a condition that is permanent for many of the lower vertebrates. Such fibres have also been described in certain muscles of some species of mammals when fully grown.

A **muscle,** in the ordinary sense of the term, is an organ consisting substantially of a mass of striped muscular fibre and its associated skeletal structures. It will therefore be convenient in this case, as in those of the cartilages and the bones, to describe the histological anatomy of the muscles in connection with their single characteristic tissue.

When seen with the naked eye a muscle appears to be made up of readily distinguishable fibres of varying fineness: these, which are the anatomical fibres, are bundles or **fasciculi** of the elements which are termed fibres in the

histological sense. Each fasciculus is invested by a layer
of areolar tissue continuous upon its outer surface with
that of those adjacent and giving off from its inner surface
delicate septa which lie between the individual fibres. The
investing layer of the fasciculus is known as the **perimysium**; the internal skeletal tissue as the **endomysium.**
The aggregated fasciculi which make up the body of the
muscle are invested as a whole by a layer of connective
tissue continuous with the outer perimysial layers, termed
the **epimysium.** The arteries and veins proper to the muscle are chiefly located in the perimysium, while the capillary network, whose meshes are as a rule greatly elongated in the direction of the fasculi, are situated in the endomysium in such a manner as to be in close proximity to
every fibre.

The striped muscular fibres terminate by rounded or
obliquely truncated ends, which are closely applied to the
correspondingly shaped extremities of white fibre bundles,
the sarcolemma of the muscular fibre being directly continuous therewith: these bundles are in some cases almost
immediately connected with the periosteum of a bone or
some similar place of attachment: in other instances they
are prolonged beyond the muscle in a fibrous mass, the
tendon of origin or of insertion, as the case may be.

CHAPTER IX.

THE SMALL VESSELS.

———

Mention was made in the introductory chapter of the fact that certain tissue aggregates, while themselves deserving to rank as organs, sustain the same relation to larger and more complex organs as do the tissues themselves. Among these compound factors of structure, as they were there termed, the most important are the small blood and lymph vessels, particularly the former. As they are built up of endothelial, skeletal and muscular tissues, their structure may now properly be described. The following statements apply, however, only to those smaller vessels which enter the structure of other organs. The larger vascular trunks will be described in a subsequent chapter in connection with the other organs of the circulatory system.

The blood vessels are commonly distinguished as **arteries,** which carry the blood from the heart, **veins,** which return it to the heart, and **capillaries,** which intervene between the arteries and the veins, and in which the blood is brought into the closest proximity to the tissues possible in a closed system of vessels. The lymph vessels originate in the interstitial spaces of the tissues (chiefly if not solely the connective tissues), these communicating directly with the open mouths of very small and thin-walled

vessels known as the **lymph capillaries**: these unite to
to form the larger vessels sometimes called lymph veins
since they convey their contents toward the heart, but
more commonly spoken of as **lymphatics.**

The interior of a **small artery,** such, for example, as can
be just distinguished with the unaided eye, is lined with a
layer of **endothelium,** whose cells are as a rule greatly
elongated in the direction of the vessel; the nuclei also
being elongated. Beneath this is a layer of elastic tissue
usually in the form of a membrane, either homogeneous
or fenestrated, but occasionally composed of reticulated
fibres. The endothelium and the elastic layer make up
what is usually called the **intima,** or inner coat; the term
is, however, applied by some histologists to the elastic
layer, to the exclusion of the endothelium.

Beneath the intima is the middle coat, or **media;** this,
in the vessels under consideration, consists almost entirely
of smooth muscular fibres, arranged in a layer several cells
deep, the long axes of the fibres crossing the vessel at
right angles or nearly so. Like all layers of muscular tissue, this is highly elastic; during life it is always upon the
stretch; and the contraction which usually takes place in
it after death throws the intima into longitudinal folds,
which, when seen in cross section, give an undulating outline to the interior of the artery which is highly characteristic.

External to the media is the **adventitia,** or outer coat:
this consists in the smaller arteries of a layer of connective
tissue in most cases clearly definable on the one hand as

pertaining to the artery, on the other passing over more
or less gradually into the adjacent interstitial tissue.

As the small arteries divide and subdivide, finally becoming lost in the capillaries, there is a gradual reduction
alike of the adventitia, the media and the intima. The former becomes reduced to a layer of extreme thinness: the
media diminishes until it is reduced to a single layer of
transverse smooth muscular fibres and later to scattered
fibres not in contact and not forming a continuous layer:
the elastic layer of the intima is similarly reduced in extent, and finally disappears; the last and least of the vessels that may with propriety be called arterial consisting
merely of the endothelial lining and an imperfectly continuous layer of smooth muscular fibres, surrounded more or
less definitely by a few branched connective tissue corpuscles.

The **capillaries** are the direct continuation of the arteries, arising either by the farther subdivision of the structures just described or springing directly, as in some cases,
from the sides of vessels still distinctly arterial in their
structure. In either case they branch freely, forming a network, whose meshes have a size, form and direction in
direct relation with the structure of the organ in which
they occur. They do not, like the arteries, become smaller
as they branch, those of any one network being approximately of the same size, though they may vary considerably in different parts of the same organ. Their size in
life is not easily determined, but the majority of them are
probably not over ten micra in diameter, though in some

tissues, notably in the marrow, they may be as much as twice as large.

⁻ In structure the capillaries are simple tubes (usually cylindrical in form) of elongated endothelial cells of such width that from two to four may be seen in the cross section of a single capillary. As in other endothelial layers, the cells are united by an intercellular cement substance : here and there patches of this substance may be demonstrated by the silver method which may be regarded as indicating gaps between the cells; these have been termed **stigmata;** and it is probable that such places offer favorable opportunity for that migration of the leucocytes from the blood-stream into the tissues which is known to be normal to them. Capillaries are almost always situated in interstitial areolar tissue or its equivalent, and are in direct relation with the connective tissue corpuscles : in some organs branched corpuscles appear to form an almost continuous layer investing the capillaries : such a layer has been termed an **adventitia capillaris.**

As the capillaries originate by the subdivision of the arteries, so their union forms the origin of the veins. A **small vein** resembles a small artery in its general structure, its wall, like that of the latter, being distinguishable into an intima, a media, and an adventitia : the chief differences between them may be stated briefly as follows. The endothelium of the veins is, as a rule, composed of shorter and broader cells than that of the arteries, and the elastic layer of the intima is thinner : the media is very much thinner, the amount of muscular tissue being very much reduced: the adventitia, which is the principal coat of the veins, is, if anything, thicker than that of the corresponding artery, and is composed largely of fibrous tissue. The veins are as a rule of a greater diameter than the arteries which they accompany.

Unlike the arteries, the veins are not during life continually on the stretch; and the recoil mentioned in connection with the former vessels does not take place after death: this reaction on the part of the arterial wall tends to drive the contained blood through the capillaries and into the veins: as a rule, therefore, the arteries are empty after death, but retain their patency on account of the thickness and elasticity of their walls. The veins, on the other hand, are usually filled with blood; or if empty become collapsed and flattened on account of the thinness and inelasticity of their walls.

A small artery and a small vein may therefore usually be distinguished when seen in cross section by the following differences: The artery is circular in form: it is in most cases empty, or contains but a small amount of blood clot: the intima is thrown into folds, giving the elastic layer a sinuous contour, upon which the nuclei of the endothelial cells are often peculiarly conspicuous: the media is quite thick, consisting chiefly of numerous muscular fibres: the adventitia is comparatively weak, and the least conspicuous of the three coats. The vein may be either circular or more or less irregular in form: if the former, it is usually filled with blood clot, readily recognizable from the numerous colored corpuscles, devoid of nuclei and appearing as clear circles, and the few leucocytes with distinct nuclei:

the intima is thin and simple in contour: the media is also
thin, containing but few smooth muscular fibres: the adventitia is often the thickest and most conspicuous of the
three coats. Where the artery and vein lie side by side,
the latter is usually the larger.

A similar application of the principles of structure above
described will enable the student to interpret the different
appearances seen when a vessel is cut obliquely or longitudinally, and with practice to recognize an artery or a
vein wherever met with in a section passing through one
of the organs of the body.

The interstitial spaces of the tissues form in many cases
an irregular network, in others definite lacunae or channels for the **lymph,** a rather indefinite term applied to the
fluid originally derived from the blood and destined to be
returned to it again by way of the lymphatic vessels. The
smallest of these, the **lymph capillaries,** arise from the
lymph-spaces of the tissues in an exceedingly irregular
manner, the connective tissue corpuscles in many cases passing by an insensible transition into their endothelium.
When fully formed they consist of tubes which resemble
blood capillaries in consisting solely of endothelial cells,
but differ from those structures in the fact that they are
rarely if ever cylindrical, being exceedingly irregular in
form, though usually flattened, and very often quite variable in diameter: they anastomose freely, forming networks of very irregular meshes. The cells of which they
are composed are usually about as long as broad, but are
usually characterized by a distinct sinuosity of outline.

The lymph capillaries unite to form the larger vessels known as the **lymphatics**. These do not differ materially from the former except in size and in the presence of valvular folds. Like the capillaries, they consist almost entirely of endothelial cells, which are, however, more elongated in the direction of the vessel than those forming the capillaries, but having the same sinuous outline. It should be stated here that by some histologists it is held that the characteristic irregular outlines seen in preparations of the lymphatic endothelium by the silver method are due solely to shrinkage of the tissues after death, the outlines of the cells in the living tissues being far simpler.

A lymphatic when seen in cross section in the interior of an organ appears as an opening of irregular form, its sides usually approaching each other (due to the flatness of the vessel); it can in many cases hardly be distinguished from a simple tear or fissure in the connective tissue, save by the definiteness of its outline: this, except in cases where the vessel is quite collapsed, is made up of simple curves; and in it may occasionally be seen the nuclei of the endothelial cells of which it is composed. In some cases the opening is larger and more nearly equal in its various diameters, resembling the cross section of a vein: it can then be distinguished from the latter by the greater thinness and simplicity of its walls and by the fact that any clot which may be present consists almost entirely of fibrin, appearing granular in the section, with here and there a few lymphocytes with large nuclei and a small amount of surrounding protoplasm.

The structure of the larger lymphatic vessels will be described in connection with the circulatory system.

The lymphatics have a distribution closely related to that of the blood vessels: and it is not unusual to see a small artery, its companion vein, and one or more lymphatics in close proximity. In some cases a blood vessel may be situated within a lymphatic vessel. Such lymphatics are termed **perivascular.**

The serous cavities of the body are in reality enlarged lymph cavities. They are lined by the **serous membranes,** which are known by distinctive names (e. g., pleura), according to their position. These consist in each case of connective tissue which contains a more or less well developed network of fine elastic fibres, surmounted by a homogeneous basement membrane. On this rests the **serous endothelium,** a layer of polygonal pavement cells of varying size. Here and there may be found small openings called **stomata,** which put the serous cavities in communication with the lymph channels of the membrane and thus with the lymphatic system: they are surrounded by cells which are usually more tumid and granular than those adjacent.

In addition to the tumid cells which surround the stomata there may sometimes be found on the serous surfaces patches of granular cells, which may be cuboidal if not columnar in form: these give evidence of rapid cell division, and may be regarded as local centres for the formation of the leucocytes which may be found in the serous cavities.

CHAPTER X.

THE NERVOUS TISSUES.

— —

The nervous tissues have for their characteristic function the reception, conduction, distribution, and discharge of impulses or stimuli. The stimuli received may be such as, when manifested in consciousness, we know as sensations ; or they may be such as are never reported to our consciousness : in either case they may be either mechanical, physical, or chemical in their origin : the distance to which they are transmitted may vary greatly in different cases, as may also the extent and nature of their distribution. The impulses which are discharged (with or without previous distribution) must be regarded as liberations of energy due to changes set up in the central nervous elements by the ingoing stimuli : they may or may not result in associated and more conspicuous liberations of energy in the muscular tissues.

The nervous elements, are, then, like the muscular elements, reservoirs of energy : they differ from them both functionally and structurally by what has been termed their polarity, by which is expressed the fact that the impulses received enter in a more or less definite direction, while the ensuing discharge takes place along a definite line which is in a general way in an opposite direction to that along which the incoming stimulus enters. A nervous element of the simplest type may be conceived of as comprising a spheroidal or spindle-shaped mass of protoplasm provided with a nucleus, each of whose extremities is continued into a filament of more or less length. Along one of these filaments stimuli travel toward the central corpuscle, and the extremity of the filament is modified for their reception: along the other, nervous impulses pass outward to their appropriate place of discharge, its extremity showing corresponding modification. The whole structure, from one set of terminals to the other (and including both) is the result of the modification of a single embryonic cell: on account of the delicacy of the various parts, however, and the great length in many cases of the conducting filament, it is rarely (if ever) possible to isolate such a nervous element in its entirety, as we can isolate a smooth muscular fibre or an epithelial cell: nor can nervous elements (with a few exceptions) be so prepared that they can be seen in their entirety in a single section: in most cases our study must at any one time be chiefly directed to one or the other of the regions above indicated. Practically, it has been customary to distinguish, as the structural factors of the nervous system, the conducting strands of greater or less length along which impulses are transmitted; the central corpuscular masses; and the terminals of reception and discharge: and although these are now known not to be independent parts, it is convenient to retain this distinction as a basis for their description: according, therefore, to the form and associated function of the various components of the nervous system, it is customary to classify them as follows.

A. Conducting elements, or **nerve fibres**: these consist in every case of continuous cylindrical (or rounded) strands which run without interruption from the central corpuscles to the terminal modifications of their free extremities. The presence or absence of certain investing structures distinguishes two kinds of nerve fibres, known as

1. **Non-medullated fibres**: also called **gray** or **gelatinous fibres**, or (from their discoverer) the **fibres of Remak**: the axis consists of a grayish cylinder or a band with rounded edges, showing delicate longitudinal striations: they are invested, if at all, by a very delicate **primitive sheath** whose existence is questioned by many histologists. Upon the surface of the fibre are seen numerous nuclei. The gray fibres not unfrequently branch and anastomose with adjacent fibres.

2. **Medullated fibres**: also called **white fibres**: the *axis-cylinder* is grayish and longitudinally striated: it is surrounded by a layer of a fatty substance termed **myelin**, the layer being called the **medullary sheath** or **white substance of Schwann**: outside of this is the clearly defined **primitive sheath** or **neurilemma**. The medullary sheath is interrupted at frequent intervals, the points where this occurs being termed the **nodes of Ranvier**; the intervening segments are called **internodes**: each exhibits one or two so-called **nerve-fibre corpuscles** situated between the medullary sheath and the neurilemma and consisting of oval nuclei sparingly invested with protoplasm. Medullated fibres rarely branch, except near their extremities.

B. Central elements, or **nerve corpuscles**: these are variously shaped bodies with conspicuous nuclei: they vary greatly in size also, some being among the smallest and others among the largest of the tissue elements. Their surfaces always give off one or more processes termed **poles**: according to the number of these processes they are commonly distinguished as **unipolar, bipolar,** and **multipolar corpuscles.** Most nerve corpuscles (in the higher vertebrates, at least) are multipolar; and in the great majority of instances one of the poles of a multipolar corpuscle can be distinguished from the others as the **process of Deiters,** or, better, as its **axis-cylinder process**: the rest are then known as **protoplasmic processes** and rapidly divide into irregular branches termed **dendrites**: in some multipolar corpuscles no axis-cylinder process can be distinguished; such corpuscles are termed **amacrine.** Other features will be discussed in a subsequent paragraph.

C. Peripheral elements, or **nerve terminals**, formed by the modification of the extremities of nerve fibres with or without epithelial or skeletal elements. According as they are situated at the extremities of fibres which conduct impulses toward the nerve centres (afferent fibres) or away from them (efferent fibres) they are distinguished as

1. **Receiving terminals**: of these various kinds are known, composing two groups which differ from each other alike in the form and in the arrangement of the structures included. In one group large numbers of similar terminals are associated with modified epithelial cells to form the essential structures of the three great organs of special sense, the nose,

§.] Peripheral elements or nerve terminals (*continued*). the eye, and the ear: these will be described in a subsequent chapter, together with the less highly specialized organs of taste. In the other the terminals are always either solitary or in groups of two or three, though they may be more or less abundant. Some of these forms of terminals have been regarded as associated with specific modes of sensation, but the function of most of them is altogether unknown: they are therefore usually distinguished by peculiarities of form or of location, or by the names of their discoverers. The principal forms may conveniently be described here.

a. The simplest form of receiving terminal is seen in what are known as **free endings**. In these the nerve fibre first looses the investing white substance of Schwann, and later the neurilemma, though the nerve-fibre corpuscles are still seen for a while before the axis-cylinder becomes entirely naked. When the latter stage is reached, the fibrils are rapidly split up into small bundles, and finally form a tuft or pencil of delicate more or less varicose filaments which ramify among the cells of the epithelium in the case of the skin, a mucous membrane, or the substance of a gland; or among the elements of other organs. Branching terminals, essentially similar, found in tendons, are known as **organs of Golgi**.

b. Closely allied to the preceding are the terminals associated with what are known as **tactile cells**: these are spheroidal cells found in the deeper portion of the epidermis and probably epithelial in character. The nerve fibre in relation with a group

C. Peripheral elements or nerve terminals (*continued*). of them looses its investments and breaks up into a number of slender branches. each of which ends in a saucer-shaped disk, the **tactile meniscus**, which embraces the proximal surface of the tactile cell, so called. There is no positive evidence that these, rather than other cutaneous terminals, are the special organs of touch.

c. The bodies known as **compound tactile cells**, also sometimes called the **corpuscles of Grandry**, found in the skin of certain birds, and also described as occurring in mammals, may be regarded as composed of two or more layers of tactile cells, and having between adjacent layers terminal expansions known as **tactile disks**. The neurilemma of the nerve fibre of which the tactile disks are the termination becomes continuous with a connective tissue capsule that invests the whole structure.

d. The name of **tactile corpuscles** has long been given to large ovoid bodies found in the papillae of the skin of the hand and foot and in other places where the sense of touch is well developed : they are also called the **corpuscles of Meissner**. Each is composed of a mass of connective tissue about which a medullated fibre winds spirally once or twice, the sheaths of the fibre then becoming merged in the mass or continuous with its capsule: the axis-cylinder passes as a nonmedullated fibre into the interior, where it branches more or less freely, the branches becoming varicose. Two, three, or even four medullated fibres may be connected with a single large corpuscle of Meissner.

e. What are termed **end-bulbs** are spheroidal or cylindroidal bodies of simple structure in which the axis cylinder of a medullated fibre enters the proxi-

C. Peripheral elements or nerve terminals (*continued*). mal end of a mass of connective tissue which is continuous with the sheaths of the fibre, and extends throughout its length, branching but little if at all. They are found in the conjunctiva of the eye, and in other modified dermal structures, as well as internally. The **articular corpuscles** found in the vicinity of joints, as well as the **genital corpuscles** of the male and the female sexual organs are probably to be regarded as modified end-bulbs, though both are held by some histologists to approach more nearly in structure to the corpuscles of Meissner.

f. The **Pacinian bodies**, or, as they are sometimes called, the **corpuscles of Vater**, are the largest of the terminals, being easily visible to the naked eye in many cases. They are irregularly ovoid in form, each having for its axis the axis-cylinder of a medullated nerve-fibre, which may terminate near the extremity of the corpuscle by a bulbous enlargement or may divide near the end into short irregular branches with pyriform extremities. The axial structure is imbedded in a cylindrical core of doubtful nature : it is faintly granular and contains scattered nuclei, and is possibly homologous with the principal mass of an end-bulb. Surrounding this is a series of concentric tunics which may be fifty or more in number, each consisting of a fibrous layer and an endothelial investment : the whole may be regarded as a highly specialized modification of the capsule of the end-bulb. Pacinian bodies are found widely distributed throughout the body. As is the case with the other forms of receiving terminals, we have as yet no certain knowledge of their specific function.

C. Peripheral elements or nerve terminals (*continued*).

2. **Discharging terminals**: Impulses sent out from
the nerve centres may give rise to muscular, glandular, or other activities. The terminals by which
discharge is made upon the elements of the tissues
involved have in every case, as far as known, the
form of ramifications or arborizations of the extremities of efferent fibres. In the case of glands
or other secretory structures the terminal subdivisions are situated among the epithelial cells. In
the case of smooth muscular tissue, nonmedullated
fibres from an adjacent ganglion or ganglionic
plexus enter the muscular layer, between whose
elements the ramifications of the fibre are situated.

The mechanism of discharge in the case of striped
muscular tissue is somewhat more complex. Medullated fibres from the intramuscular plexus, following the endomysium, divide in each instance
into two or more branches, each branch passing
to a single muscular fibre. The medullary sheath
disappears, the neurilemma apparently becomes
continuous with the sarcolemma of the muscular
fibre, and the axis-cylinder breaks up into a
number of fine varicose branches: the latter rest
upon or are imbedded in the **sole-plate**, a flattened granular mass of protoplasm containing
several nuclei and lying between the sarcolemma
and the body of the fibre: the whole structure is
termed a **motor end-plate**. By some histologists
the end-plates are believed to be situated altogether
outside of the sarcolemma.

Attention has been called to the delicate longitudinal striation of the axis cylinder of the medullated fibre and of the corresponding portion of the gray fibre. This is the expression of a distinct fibrillation, the **primitive fibrillae** being imbedded in an intervening homogeneous substance, the **neuroplasm** (the resemblance to the fibrillae and sarcoplasm of striated muscular fibre is noteworthy). The terminal subdivision common to medullated and nonmedullated fibres consists of a breaking up of the axis into smaller bundles of these fibrillae and eventually, in some cases, to the separation of each individual fibrilla.

The gray or nonmedullated fibres consist of little more than bundles of fibrils and neuroplasm. Those which compose the branches of the olfactory nerves have a well-defined and nucleated primitive sheath: in most gray fibres no such sheath can be demonstrated. Scattered nuclei are seen upon the surface of the gray fibres: these, like the sheath, when present must be regarded as skeletal rather than nervous in character. Gray fibres show well marked varicosities, which are possibly due to local accumulations of neuroplasm, the fibrillae being correspondingly separated at such points. Where gray fibres branch and anastomose bundles of fibrillae accompanied by neuroplasm pass over from one fibre to another: the angles formed by the branches are frequently filled for a short distance with neuroplasm, and investing nuclei are often relatively abundant at such points.

The axis cylinders of white fibres show under certain methods of treatment transverse striation curiously like that of a striped muscular fibre: if a reality, and not due merely to the reagents used for its demonstration, this must be caused by regular variations in the size of the fibrillae. The axis cylinder may also be shown to be suddenly thickened at the nodes of Ranvier, the spindle-shaped enlargement being apparently due to an increase in the amount of neuroplasm with an accompanying separation of the fibrillae. Some histologists maintain that the axis cylinder is invested with a delicate structureless sheath, for which Kuehne has proposed the name of the **axilemma**: whether this structure exists in the living fibre is still a matter of question.

The medullary sheath also presents (under certain treatment) apparent evidence of a structure that would hardly be inferred from its semifluid character as seen in the fresh nerve. A reticular framework of a substance of a horny nature known as **neurokeratin** can be demonstrated, whose meshes and filaments vary greatly in size in different parts of the same fibre. That the substance in question exists as a component of myelin is probably true: but the solid framework described is quite possibly due to its coagulation by the reagents employed. Far more conspicuous are the oblique clefts seen in the medullary sheath after treatment with certain reagents, notably osmic acid: these are evidently the view in section of conical cleavage spaces running from the primitive sheath to the axis cylinder, and dividing the medullary sheath into the **medullary segments** of Schmidt and Lantermann, a number of which may be found in each internode: whether these are real or artificial must, however, be regarded as still unsettled.

The neurilemma exhibits no special structural features worthy of remark. It should be noted that when a medullated fibre joins the brain or cord, while the medullary sheath is continued within the axial structure as far as the gray matter, the neurilemma disappears; thus the columns of the cord are made up in great measure of medullated fibres devoid of neurilemma. At the distal extremities of the fibres the medullary sheath is the first to disappear, the neurilemma being continued for some distance toward the terminal.

Nerve fibres, both gray and medullated, vary considerably in size, their diameters ranging from two to twenty micra; the difference appears to be associated with a corresponding difference in the length of the fibres.

Corpuscles, fibres, and terminals are now known to be continuous structures and components of what may properly be called true tissue elements, meaning by that term in each case the result of the modification of a single embryonic cell. As indicated at the outset, such an element may consist of a receiving terminal, an afferent fibre (medullated or nonmedullated), a central corpuscle, an efferent fibre (of either kind) and a discharging terminal. The simplest form of terminal is in either case a tuft of fibrillae: if the subdivisions of the receiving terminal are called **dendrites**, and the discharging cluster an **arborization**, the two can readily be distinguished by these terms. A corpuscle so situated would be essentially **bipolar**; such corpuscles exist, though not in great numbers, in the nervous tissues of the higher vertebrates; more frequently the points of attachment of the two fibres become approx-

imated and finally consolidated for a short distance, forming what is apparently a **unipolar** corpuscle with what is
termed either a Y- or a T-connection according to the
mode of separation of the two fibres. In certain super-
ficially situated elements of a sensory character in some of
the lower animals (and possibly in higher forms as well)
the receiving terminal and afferent filament become so
shortened and condensed as to form a mere eminence only
on the body of the corpuscle: such elements may be said
to be in form (but even then not in function) **unipolar.**
What have been called in the past **apolar** corpuscles probably do not exist.

In the ganglia of the sympathetic system corpuscles are
found with more than two processes, each of which becomes an axis cylinder (or a gray fibre): such corpuscles
are in the strictest sense **multipolar**: whether the majority
of the poles are afferent or efferent is unknown: both
conditions may possibly occur.

The term **multipolar** has long been applied to the corpuscles found chiefly in the brain and spinal cord in which
a distinction can be made, as has been pointed out, between a single **axis-cylinder process** and a number of so
called **protoplasmic processes** which subdivide into a
group of **dendrites.** It has been suggested that the latter
have for their function some connection with the nutrition
of the corpuscle: but a more reasonable interpretation is
one which regards such a corpuscle as resulting from the
disappearance of the afferent fibre, its primary subdivisions thus becoming processes of the corpuscle itself. The
axis cylinder process may in its course give off one or more

slender branches; these leave the process at well marked
angles, but soon after bend strongly to become approximately parallel to it in most cases: they are known as
collaterals, and like the processes from which they arise
terminate in **arborizations**.

Axis-cylinder processes which pass from the gray into the
white matter of the cord become invested with a medullary sheath and are then true axis cylinders: elements in
which this is the case are known as **corpuscles of the first
type**: in other cases the efferent process is quite short, the
terminal arborization being situated in the gray matter:
such elements are called **corpuscles of the second type**.
The disappearance of the process altogether, making the
arborization sessile, like the dendrites, gives rise to the
amacrine corpuscles of Cajal.

Nerve corpuscles always have large and conspicuous
nuclei, in the vicinity of which a patch of pigment granules
is very commonly present. The fibrillae of the processes
may be traced into the interior of the corpuscles, but their
internal distribution is as yet unknown. The corpuscles
are almost always situated in well defined lymph spaces
which agree closely with them in contour. The forms of
the corpuscles of the brain and cord will be described in
the chapter devoted to those organs.

The nerves are definite aggregates of nerve fibres: like
the blood vessels, they penetrate the organs of the body
and are consequently to be regarded among the factors of
structure thereof: the same is true of many ganglia: both
will therefore be described at this time.

A **nerve** is a bundle of nerve fibres or an aggregate of such bundles. Each bundle is termed a **funiculus**, and is composed of a number of fibres surrounded by a cylindrical sheath called the **perineurium**. The latter is lamellated in structure, the number of lamellae never being less than three save in the smallest branches of the nerves: they are separated by distinct lymph spaces lined with endothelioid corpuscles. The inner lamella is continued into the funiculus by the connective tissue which lies between the fibres and supports their capillaries, called the **endoneurium**. In small funiculi this connective tissue is homogeneous in composition, approaching gelatinous tissue in consistency : such funiculi are termed **simple**: larger funiculi, called **compound**, show here and there in the endoneurium connective tissue septa which divide the funiculus irregularly.

In small nerves, consisting of but a single funiculus, the outer lamella of the perineurium is continuous with the adjacent areolar tissue : where several or more bundles are associated, however, as in the larger nerves, a definite mass of connective tissue, containing more or less fat, and definitely compacted on its outer surface, invests and supports the funiculi, becoming continuous with their outer lamellae: this is known as the **epineurium**. Within it the associated funiculi divide and anastomose from time to time, each large nerve being thus in reality a greatly elongated plexus.

As the funiculi divide into small groups of fibres and finally into single fibres in the vicinity of their destination, the perineurium becomes greatly reduced, being finally continued for a short distance on the single fibres either as a single lamella or as a mere layer of endothelioid cells: such an investment is known by the name of **Henle's sheath**.

A **ganglion** is a mass of nerve corpuscles invested with a definite sheath or capsule of connective tissue continuous with the epineurium of the nerves with which it is associated; or, in the case of nerves consisting of single funiculi, with the perineurium. These may be but two in number, the ganglion in such cases being practically seated upon a nerve trunk; or there may be three or more, the ganglion being situated at their intersection. In almost all the larger ganglia there may be clearly distinguished a cortical portion, consisting chiefly of nerve corpuscles, and a central portion, consisting largely of nerve fibres. Of these some pass directly through the ganglion, while others pass into or out from the cortical portion, being connected with the corpuscles. Each corpuscle is, as a rule, contained in a delicate capsule continuous with the neurilemma of the associated fibre or fibres and enclosing, as has already been stated, a pericorpuscular lymph space.

The **neuroglia**, or sustentacular tissue of the brain and cord, has been referred to in the chapter devoted to the fibrous tissues. It differs from all the tissues of that group in its origin and in the absence of anything like the matrix characteristic of them; consisting, as was stated, entirely of peculiar branched corpuscles known as **glia-cells**. These are stellate or irregularly shaped cells with large nuclei, which stain conspicuously with some reagents. Their branches, which are quite numerous, terminate in long slender processes: these are stated by Ranvier to be fibrillated, the fibrillae passing through the body of the cell from one process to another: they are variously arranged upon glia-cells from different parts of the cerebrospinal axis, those of the gray matter of the cord, for example, differing from those of the white in the number and disposition of their processes. In a general way it may be said that the latter form a dense reticulum of closely interwoven fibres: this in sections has a finely granular appearance conspicuously seen in the gray matter..

The glia-cells are found in the cerebrospinal axis and are derived from the same embryonic layer as the nervous elements themselves. While, therefore, their function is probably purely mechanical, or, in a sense, skeletal, they must be regarded as closely related to the nervous tissues rather than to the skeletal tissues proper. These latter also penetrate the cerebrospinal axis, in the form of connective-tissue trabeculae which compose a proper skeletal framework: and although they gradually diminish by subdivision and become reduced to delicate fibrils intermingled with those of the neuroglia, nothing like a transition from one tissue to the other has ever been observed.

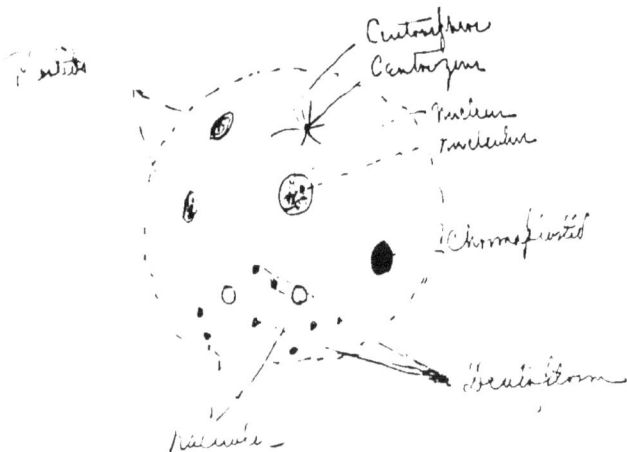

CHAPTER XI.

THE STRUCTURE OF THE CELL.

We have now passed briefly in review the various tissue elements, considering both their form and characters and their union to compose the tissues of the body; as also the structure of some of the simpler aggregates of tissues, or organs. The elements of the tissues were at the outset defined as cells or as derived from the modification of cells; and a cell was defined as a nucleated mass of protoplasm. It is important now for us to consider the structure of the protoplasm itself, and of the nucleus as well; and to learn something of the process by means of which new cells are formed.

It was stated in the opening chapter that the **protoplasm** which makes up the body of the cell is neither homogeneous or structureless, as it was once supposed to be. The delicate granulation generally characteristic of its appearance as seen by ordinary powers proves with more improved means of research to be the expression of a delicate **reticulum** or network of a somewhat denser substance which has been given the name of **spongioplasm**; its meshes are filled with a less dense or semifluid substance designated as **hyaloplasm**: the proportion between the amounts of these two substances may vary greatly: as a

general rule the relative amount of the former increases
with the age of the cell. The meshes of the spongioplasm
may vary greatly in size and in the coarseness or fineness
of their constituent fibrils: the accumulations at their in-
tersections are the granules most readily seen.

In cells which become surrounded by a cell wall composed of some formed product (e. g., an epidermal cell),
the reticulum becomes quite close and dense near the surface: but in many cases (e. g., a leucocyte) the converse is
the case, the exterior of the cell consisting almost, if not
quite wholly, of hyaloplasm. Such a clear outer portion is
sometimes termed **ectoplasm**, in distinction from the granular inner portion known as **endoplasm**: the distinction
is, however, of questionable value, since in some of the
lowest animals the conditions are reversed, the same
terms being applied to a denser outer and a more fluid
inner portion of the cell. The terms **paraplasm** and **deu-
toplasm** are also sometimes made use of in connection
with the structure of the cell body to designate granules
imbedded in the protoplasm, and consisting either of substances taken up by the protoplasm in a solid form, or of
formed products temporarily stored in the cell; such, for
example, as yolk granules in the ovum.

The **nucleus** gives evidence of a reticular structure even
with ordinary powers; and this structure is also clearly
seen to vary in the different nuclei of adjacent cells or in the
same cell at different times if watched while still living.
Under ordinary conditions, however, the nucleus when
seen in what is usually designated the "resting" condi-

tion is a spheroidal vesicular body bounded by a well defined wall, and containing the network above referred to: this is sometimes fine and close meshed; at others composed of but a few coarse fibrils, which are in some cases quite irregularly disposed; in some cases the fibrils form a continuous filament which is arranged in a tangled skein. The nodes of the network in many cases form coarse granules, which are quite conspicuous: in addition there are often seen in the nuclei distinct spheroidal bodies apparently different in composition from the network: such a body is called a **nucleolus**.

The meshes of the network are filled with a clear semi-fluid substance which is not readily colored by the staining fluids which render the network and nucleoli conspicuous. This difference between the two principal substances of the nucleus led Flemming to propose for the substance composing the filaments the name of chromatin, and for the clear substance that of achromatin. More recent researches have made clear, however, the fact that the denser portion of the nucleus itself consists in part of a substance which does not stain any more readily than the more fluid portion: while the latter, now usually termed the **nuclear matrix**, is regarded as possibly similar in composition to the hyaloplasm of the cell-body. The name of **chromoplasm** has therefore been proposed by Carnoy for the substance which forms the filaments, that portion which is readily stained being designated by the term **chromatin**, and that which resists ordinary stains by the term **achromatin**; a different application of these terms from that originally proposed by Flemming.

The nuclear wall is composed of chromoplasm: by some
it is regarded as consisting merely of a fine and close network of that substance; by others as a definite and continuous layer. Its continuity with the network is evident
from its comportment at the time of nuclear division.
Whether the nucleoli consist of chromoplasm must for the
present be regarded as an open question.

While the arrangement of the filaments of the nuclear
network is often exceedingly irregular, especially in the
resting stage, it can often if not always be seen at the
time of its greatest development to have a definite plan,
whose basis is the formation of a larger or smaller number of elongated **loops** having a meridian-like arrangement in relation to a definite axis. The turns of the loops
are directed toward one end of the axis, termed the **pole**
of the nucleus: the free extremities meet (and frequently
interdigitate) in the region around the opposite end of the
axis, the **anti-pole**. The sides of the loops are often exceedingly irregular in their course, and may in addition
branch frequently, the branches anastomosing and thus
forming the irregular network commonly seen. Where
the ends of adjacent loops become continuous the convoluted filament sometimes seen is produced.

The division of older cells to form new ones is preceded
in man and the higher animals generally by **nuclear
division**. The older observers (whose imperfect microscopes showed them in the interior of the nucleus only the
nucleoli and the coarser nodal granules) believed this process to be quite simple in its nature, consisting merely in

the passage of a cleavage-plane through the nucleus in the same way as is seen in the cell body itself. This, which is called **direct division**, may possibly sometimes occur; but the constant advance of our knowledge makes it yearly more and more evident that the common mode of nuclear division as it occurs in plants and animals, and in normal and pathological changes alike, is an exceedingly complex process, to which the rather unfortunate name of **indirect division** is commonly applied: and it is quite possible, if not probable, that this is the sole method. The series of changes involved in this process has been termed **karyokinesis.** The name of **mitotic division** has also been applied to the process, the successive stages being called **mitoses**: the so-called direct division being distinguished as **amitotic**.

The successive steps that can be recognized in what is really one continuous process have been designated by special names, based on the appearances presented by the nucleus from time to time. The first step is the formation of what has been variously termed the **spirem**, or **close skein**: the secondary filaments are retracted into the primary filaments or loops; the nuclear membrane is also absorbed, as are the nucleoli; the latter fact indicating the possible identity of these bodies with the chromoplasm of the network. The next step is the formation of the **open skein**, or **wreath**: the primary filaments contract, becoming shorter and stouter, and having a less tortuous course: they gradually assume the form of an equatorial wreath of loops with, as has been stated, their flexures turned toward the region of the pole. To these loops the name of **chromosomes** has been given.

While the·chromosomes are thus being defined, there appears in the polar area a group of fibres of achromatin, known from its form by the name of the **achromatic spindle**: this gradually moves toward the centre of the nucleus, taking an axial position with its extremities directed toward the pole and the anti-pole respectively. The chromosomes having by this time become quite short and stout, and V-shaped from the divergence of the limbs of the loops, attach themselves to the spindle, eventually assuming a radial position at its equator. When the nucleus is viewed at this stage from a polar or an anti-polar direction the radiating arms of the chromosomes together form a starlike figure to which the name of the **aster** has been given.

During the formation of the aster or immediately thereafter, the **cleavage of the chromosomes** take place, each loop being split into two similar (but of course more slender) loops by a plane passing through them all equatorially. This is the central process in the division of the nucleus: the changes which follow have therefore been sometimes designated by the term **metakinesis**: they are in a certain sense the retracing of the processes already described. In its more limited sense the term metakinesis is applied only to the cleavage of the chromosomes, and the changes immediately connected therewith.

The daughter-loops formed by the cleavage of the chromosomes begin to separate first at their apices, these being turned toward the extremities of the spindle: they then travel slowly along the achromatic fibres in each direction, two sets of chromosomes thus being formed; these gradually arrange themselves about the poles of the spindle, which have now become the polar areas of the two daughter-nuclei. The limbs of associated loops remain for some time connected together by delicate achromatic **uniting filaments:** the whole figure seen from the side has a resultant barrel-shaped appearance; at either extremity the chromosomes have a stellate arrangement, and this stage is therefore designated the **dyaster**.

As the dyaster is formed, a cleavage plane passes through the body of the cell, whose course in the nuclear region is sometime marked by nodal points on the uniting filaments. As their separation is completed, the free extremities of the chromosomes of the daughter-nuclei bend inward toward the new antipolar areas, which face the plane of cleavage and therefore, also, each other. Changes now ensue in the inverse order of those described as taking place at the beginning of the process, the chromosomes becoming first converted into **open skeins,** and later into the **closed skeins** whose farther modification gives rise in each to a **reticulum** by the formation of anastomosing secondary filaments, accompanied with or followed by the formation of a nuclear membrane, the appearance in some cases of nucleoli, and the final assumption of the **resting stage.** The stage in which two adjacent and parallel skeins are seen is sometimes termed the **double skein** or **dispirem:** but it is evident that we are here dealing with structures which, though genetically associated, are now parts of separate and distinct nuclei.

The number of the chromosomes varies considerably in different plants and animals, but is probably constant for the same tissues in each species: their form may also vary greatly, particularly as regards the length of the branches of the loops. The changes above described may in many cases be followed step by step with a good microscope of ordinary powers, either as they take place in living cells, or as they may be found in adjacent cells of suitably prepared tissues, each of the characteristic figures mentioned being clearly recognizable: in some instances, however, owing to the irregularity in the form of the chromosomes, or to variations in the rate of their transformation, some of the phases may be so far modified as to be no longer distinguishable. The essential features of the process are, nevertheless, always to be discerned, and should be distinctly borne in mind: they are, in their order, as follows: first, the collection of the chromatin of the nuclear network into chromosomes; second, the equatorial arrangement of the latter in what has been termed the **nuclear plate**; third, their metakinetic cleavage; fourth, the separation of the two sets of chromosomes thusformed; and finally their resolution into the nuclear networks of the two resultant daughter-nuclei, whose formation by this method is accompanied by the cleavage of the protoplasmic body of the parent cell, thus completing the formation of new cells.

It has recently been made fully evident that the changes taking place in the chromatin of the nucleus during karyokinesis are accompanied and preceded by other equally complicated changes in structures made up wholly of achromatin, the formation of the achromatic spindle being a portion thereof. While these changes are doubtless as important as those already described, their nature is as yet far less clearly understood. A description of them will therefore not be necessary at this time.

END OF PART I.

PART II.

HISTOLOGICAL ANATOMY.

CHAPTER XII.

INTRODUCTORY.

—

Histological Anatomy has already been defined as the study of the arrangement of the tissues to form the organs of the body: an **organ** has also been defined as a particular part of the body having a definite form and function; and it is a familiar fact that organs having common or essentially similar functions are associated together under the name of a **system,** whether they are continuous, as in the case of the nervous, or discontinuous, as in that of the muscular system.

While the study of the histological anatomy of the organs can in most cases be pursued most naturally by considering them in their relations as components of the various physiological systems, on account of the community of structure which, in most cases, characterizes associated organs, and while structure and function are without question closely (though not always evidently) related, it should always be kept clearly in mind that we are here concerned with structure only; and particularly with structure as composed of tissues: our constant endeavor should be, in the first place, to analyze the organs into the tissues of which they are composed, and to determine the relations of each to the others; and in the second place to note any characteristic peculiarities exhibited by any of

the tissues present; this should be accompanied in each instance by a careful consideration of the disposition and
characters of the compound factors of structure present,
such as the blood, lymph, and nervous supply.

Tissues have already been defined as masses of cells or
of cell-derivatives; and since tissues compose the organs,
and organs make up the whole body, it follows that the
body is to be regarded as a mass of more or less modified
cells. The innumerable cellular elements which make up
the adult human organism are in every case derived from
the division of previously existing cells, and are therefore
necessarily the descendents of a single ancestral cell. That
cell is the **fertilized ovum** or, as it is sometimes termed,
the **oosperm.** The study of fertilization, of the segmentation of the oosperm, and of the subsequent development of
the tissues and organs of the body lie strictly within the
province of the science of Embryology; but a brief statement of the origin of the various tissues may with advan-
tage be given here, as throwing light on the structure and
relations of the organs of the body.

Repeated cell division or **segmentation** in a short time
divides the oosperm into a spheroidal mass of apparently
similar cells: these soon arrange themselves in two distinct layers, from which a third intermediate layer is
shortly afterward derived: the outer of these layers is
called the **epiblast** or **ectoderm,** the middle the **mesoblast**
or **mesoderm,** and the inner the **hypoblast** or **entoderm;**
the whole trilaminar structure receiving the name of the
blastoderm.

CHAPTER XII. INTRODUCTORY.

From the cells which compose the epiblast are derived the following structures:

The epidermis, and its appendages the hairs and the nails, and the epithelium lining the tegumentary glands (sweat glands, sebaceous glands, mammary glands).

The epithelium of the nasal passages and the associated cavities and glands.

The epithelium of the mouth and of the glands continuous therewith, and of a portion of the tongue: the taste organs: the enamel of the teeth.

The epithelium of the conjunctiva and of the glands of the eyelid, and of the front of the cornea: the lens of the eye: the retina (secondarily as an outgrowth from the brain).

The epithelium of the membranous labyrinth of the ear.

The epithelium lining the cavities contained in the cerebrospinal axis: the nervous tissues: the neuroglia: the pineal body: the pituitary body.

From the cells which compose the middle layer or mesoblast are derived the following structures:

The epithelium of the urinary and genital organs (with the exceptions of the epithelium of the bladder and urethra), including the reproductive elements of both sexes.

All the muscular tissues of the body, with the exceptions of the cells (doubtfully muscular) found in the sweat glands.

The skeletal tissues of all sorts throughout the body.

The blood-vascular and lymph-vascular system : the serous membranes : the spleen and other adenoid bodies : the blood and lymph corpuscles.

It should be stated that by some histologists the cells which give rise to the tissues of the first two groups are regarded as having an origin somewhat different from those giving rise to the last two: the name mesoblast has been retained by them as a collective title for the former, while for the latter the name of **parablast** was proposed by His, and later that of **mesenchyma** by the Hertwigs. While this is probably true of the tissues of birds, it has not yet been proven for any mammal: and there are special reasons why it might be true in one case and not in the other.

From the cells which compose the hypoblast are derived the following structures:

The epithelium of the back of the tongue, the lower part of the pharynx, the oesophagus, stomach and intestines: that of all the glandular appendages of the alimentary canal.

The epithelium of the Eustachian tube and middle ear.

The epithelium of the larynx, the trachea, the bronchi, the bronchial tubes and the air sacs of the lungs.

The epithelium lining the urinary bladder and the urethra.

The epithelium lining the vesicular alveoli of the thyroid body.

The concentric corpuscles or epithelial nests of the thymus.

Even a brief study of the tabular statement of the origin of the tissues of the body above given will make clear the facts that most organs are made up of tissues derived from more than one of the primary tissue-layers, and that in some cases at least, tissues which are structurally continuous and to all appearances similar are of different embryonic origin: this is notably the case with the transitional epithelium found in the ureters and the bladder; and other instances might be mentioned.

The order of study pursued in acquiring a knowledge of the histological anatomy of the various organs and systems of the body is plainly a matter of convenience. That which will be here pursued is one shown by experience to be desirable on some accounts: but it should be understood that it may be readily varied at will. Omitting from farther consideration the simpler organs already described in the preceding part in connection with the tissues which chiefly compose them, such as the cartilages, the bones, the muscles, etc., the various regions, systems or groups of organs will be successively described as follows:

The **tegumentary system** will first receive attention: this includes not only the skin, but also those solid appendages, the hairs and the nails, which are derived from the special modifications of its outer layer; and those ingrowths of the same layer which constitute the sudoriparous, sebaceous and mammary glands.

The skin upon the outer surface of the lips is continuous with the so-called mucous membrane of its inner surface: a transition which brings us naturally to the **mouth and its contents**: this includes the study of the lining membrane above referred to; the buccal and other glands which open thereon; the teeth and the tongue.

The mouth is the antechamber of the **alimentary canal**, though it is often regarded as a part of it. Beginning with the pharynx, we naturally consider in their order the oesophagus, the stomach, the small intestine in its various regions, and the large intestine (including the rectum); and also the glandular appendages of the canal, the liver and the pancreas.

The pharynx is not only a portion of the alimentary canal, but of the **respiratory tract** as well: the latter being, as embryology shows, an outgrowth of the digestive tube; its study includes that of the larynx, the trachea and bronchi, the bronchial tubes in their various ramifications, and the terminal sacs which make up with them the proper substance of the lungs.

The bladder and the urethra are parts of an outgrowth from the posterior region of the alimentary canal, as the respiratory tract is of the anterior. With them as median structures are closely associated the paired organs which complete the **urinary apparatus**: and intimately related therewith are the male and female **reproductive glands**, and the accessory organs connected with reproduction and micturition. This group will next be studied.

Following the groups of organs above indicated, the **circulatory system** may next receive attention. This includes the study of the heart, the larger blood vessels, and the greater lymphatic trunks, the smaller blood vessels and lymphatics having been already considered. Since the great serous cavities, such as that of the thorax or of the abdomen, may best be regarded as lymph spaces, the special discussion of their lining membranes may appropriately be considered here.

In addition to the lymph nodes, there are found in the body larger organs apparently allied to them or derived from their modification, such as the spleen and the thymus. The name of **ductless glands** has long been applied to these and to other soft organs of uncertain function, such as the thyroid, the pituitary body, and the adrenals (the so-called suprarenal capsules). The group (to which the name of the **adenoid bodies** is sometimes applied) is a heterogeneous one, its members in some cases having little in common: but they may for convenience be considered together.

The structure of the **central nervous system** is at present being worked out chiefly by histological methods: it is not as yet always easy to distinguish between what may be regarded as physiological, and what as histological anatomy in some cases. Imperfect as our knowledge is at present, the briefest statement of its details would transcend the limits of an elementary course in histology; but some knowlede of itgs most salient features is essential: the structure of the spinal cord will be discussed, together with that of the principal regions of the brain. On account of their intimate relation to the cerebrospinal axis, the membranes which invest it and which form the lining of the spinal canal will be taken up in this connection.

Finally the study of the central nervous axis may properly be followed by that of the complex outlying structures which are the essential parts of the **organs of special sense,** such as the eye, the ear and the nose. Beginning our course with the common investment of the body, we close it with organs which are in a great measure specializations thereof.

CHAPTER XIII.

THE SKIN AND APPENDAGES.

The **skin,** which is the investing and protecting membrane of the body, is, like other membranes found upon free surfaces, composed of two primary layers, one of which is epithelial, the other skeletal: each of them being capable of division into more or less well marked secondary layers or strata. The outer or epithelial layer is known variously as the epidermis, cuticle, or scarf skin: the inner or skeletal layer as the derma, corium, or cutis (or less properly the cutis vera or "true" skin).

The **epidermis** can under favorable circumstances be seen even with the naked eye to be made up of two distinct layers, an inner moist or **mucous layer** (sometimes called the **rete mucosum**), and an outer dry or **horny layer;** and each of these layers can be resolved by the microscope into strata characterized by differences in the form and arrangement of the component cells: it may therefore be regarded as the best example of a stratified squamous epithelium in the whole body. Beginning with the mucous layer, we find next the corium cells which are columnar in form, and which are constantly undergoing division; the new cells formed at their free extremities are at first vertically elongated, then polyhedral in shape, and later somewhat flattened vertically: they form with the basal cells the first or lowermost stratum of the epidermis, known as the **stratum Malpighii**: the elements composing this stratum have the form of prickle-cells, their numerous short processes preventing the actual contact of the surfaces of adjacent cells, thus forming channels for the circulation of lymph and the nutrition of the elements: all the cells of this stratum may be regarded as living. Here and there leucocytes may sometimes be found; they have wandered into the epithelium from below and occupy irregular intercellular spaces. The cells of the deepest portions of the stratum Malpighii contain pigment granules, the color of the skin in different races depending chiefly on the relative abundance and color of the pigment. Delicate nerve fibrils enter this stratum of the epidermis, and, as stated in a previous chapter, Merkel has described special terminals thereto under the name of tactile cells.

As new cells are constantly formed in the deeper portions of the stratum Malpighii, the older cells are as constantly pushed farther and farther from the blood vessels of the corium which constitute their basis of nutritive supply through the agency of the lymph channels already mentioned. At a certain distance, which varies in different portions of the body, they begin to yield more conspicuously to the mechanical pressure from without, becoming more flattened in form, and at the same time to undergo degenerative changes, granules of a fat-like compound termed **eleidin** appearing in their substance in great numbers. There is thus formed a definite layer never more than a few cells deep to which the name of **stratum granulosum** is applied: the stratum granulosum and the stratum Malpighii together make up the mucous layer, and in most parts of the body the greater portion of the epidermis.

Immediately above the stratum granulosum and sharply distinguished from it is a thin layer of cells which resemble those of that stratum in being compressed in form, but differ from them in greater homogeneity of substance and therefore in translucency: this is the **stratum lucidum**, the lower of the strata of the horny layer. According to Ranvier the formation of eleidin is followed by its transformation into **keratin**, the characteristic substance of horn, nails, claws, etc., which are, as we shall see, in the main developments of the stratum lucidum.

While the stratum lucidum is constantly receiving accessions from the cells of the stratum granulosum upon its lower or inner side, it is as constantly undergoing modifications on its upper or outer surface; and the line which marks this transformation is equally well defined in either case, the stratum lucidum remaining like the stratum granulosum of nearly constant thickness and definite limitations. The cells as they pass from its outer surface become somewhat swollen and more loosely disposed, forming a layer in which the outlines of the individual cells may be clearly discerned: the nuclei have disappeared and all traces of protoplasmic structure. This layer is sometimes designated the **stratum corneum**: in places where there are no hairs upon the surface it can be divided into two layers, the lower, or **stratum epitrichium**, consisting of thicker cells, more loosely disposed, and the outer or

stratum squamosum of thin closely oppressed scales, which are eventually and constantly cast off. The stratum lucidum and stratum corneum (with the subdivisions of the latter) together make up the horny layer.

The epidermis varies greatly in thickness in different parts of the body: it may be no more than a tenth of a millimetre in thickness, or may be as much as a millimetre or more in places where the pressure and friction upon the surface is greatest, even under ordinary circumstances: and under special conditions the cells of the stratum squamosum of the palms of the hand or the soles of the feet, instead of being exfoliated, may become impacted to form layers two or three millimetres in thickness. The external surface shows irregularities which in some measure correspond to the conformation of the corium below, but as a rule differ therefrom by their less extent: the surface of the corium is, as we shall see, covered by projections known as papillae, into the spaces between which the stratum Malpighii descends: but this stratum varies correspondingly in thickness to such an extent that the papillary elevations are reduced on its upper surface to mere undulations, to which the remaining strata conform. The deeper lines upon the surface of the epidermis correspond to definite folds in the corium.

Like the epidermis, the **corium** can be divided into strata, which are not, however, so clearly defined in the case of the latter as of the former, consisting as they do in modifications of a fibrous layer. Upon the surface the fibres are fine and very closely felted, forming a thin, homogeneous and almost translucent stratum, which may be distinguished as the **basement membrane** of the epidermis. Its surface is closely beset with minute projections, which interlock with corresponding irregularities at the lower extremities of the columnar cells which lie at the base of the stratum Malpighii.

Beneath the basement membrane above referred to the corium consists of a felted mass of rather coarse bundles of white fibrous tissue, reinforced by elastic fibres in varying quantity, and containing in some localities a greater or less amount of smooth muscular fibres, notably in connection with the hair-follicles. The outer portion is denser, the bundles being smaller and more closely felted and lying in the main parallel to the surface: it bears the papillae already referred to in another connection, and has therefore been variously designated from its structure and conformation as the **dense stratum** or the **stratum papillare**: below it passes rather abruptly, but with no well-defined line of demarkation, into a region in which the bundles are coarser and less numerous, being more loosely and irregularly disposed: from its structure this is known as the **stratum reticulare.** The **papillae** are conical or club-shaped projections upward of the dense layer: in many cases they are more or less subdivided at their free extremities, in which case they are known as **compound papillae:** they are most abundant in the regions of the skin where the sense of touch is most acute, and also in a modified form make up the dermal portion of the nail-bed. In some places, notably on the tips of the fingers, they are arranged in single or double rows along ridges of the corium, forming the familiar patterns readily seen with the naked eye: these patterns remain constant for each digit of each individual throughout life, and have therefore been used as marks of identification. Where best developed the papillae are from an eighth to a fourth of a millimetre in height.

The connective tissue corpuscles of the corium, like those of other fibrous membranes, are small, and as a rule compressed, their long axes lying parallel to the direction of the bundles with which they are associated: leucocytes are also present, as are pigment cells in moderate numbers.

The meshes of the stratum reticulare not unfrequently contain clusters of fat-cells of varying extent: immediately beneath, in most portions of the body, we come to a layer of larger or smaller ovoid or polyhedral fat-lobules separated by a coarse meshwork of fibre bundles. This layer, frequently termed the **panniculus adiposus,** is often spoken of as subcutaneous: it is, however, in many cases no more clearly marked off from the stratum reticulare than is the latter from the stratum papillare: it may with good reason be regarded as a portion of the skin, and as such be distinguished as the **stratum adiposum:** whether it shall be regarded as cutaneous or subcutaneous is largely a matter of definition; the fact of its association with the other structures of the skin as a part of the temperature regulating mechanism of the body is important. Below it passes into the loose layer of **subcutaneous areolar tissue,** which, save in a very few localities, intervenes between the skin and the structures beneath, permitting of its more or less free movement upon them.

The skin, exclusive of the stratum adiposum, varies in thickness from half a millimetre to two or three millimetres: and may even occasionally be as much as twice the latter quantity in thickness. It is thickest upon the shoulders and back.

The arteries which pass to the skin branch and subdivide in the subcutaneous tissue, the small vessels thus formed proceeding toward the surface: on their way they give off twigs which supply the fat-lobes above mentioned, and the hair follicles, sweat glands, etc., presently to be described. As they approach the surface they branch and anastomose, and finally break up into a meshwork of capillaries, situated just below the basement membrane, supplying the papillae together with the other portions of the dense layer. The capillaries unite to form a superficial venous network, the larger veins arising therefrom passing to the deeper portions of the skin in such a way as to accompany the arteries in large measure.

Lymphatics arise in the spaces between the bundles of fibres which make up the dense layer. These are so disposed as to form a lymphatic network just below the superficial capillary network just mentioned: lymphatics have also been demonstrated in some of the larger papillae. A second network is said to exist in the deeper portion of the corium, the two communicating freely with each other and with the subcutaneous lymphatics.

The nerve supply of the skin varies greatly in different portions of the body. Like the blood vessels, the nerves form
plexuses in the papillary region, the meshes immediately
beneath the epidermis becoming very fine and close: from
the fibres composing them fibrillae are sent up into the
stratum Malpighii in the manner already described. As
stated in a previous chapter, it is not yet certain how
these fibrils terminate.

In certain (if not all) portions of the body, more or
fewer of the papillae are supplied with nerve fibres which
there terminate in the so-called tactile corpuscles of Meissner: these are most abundant in the papillae of the fingers
and toes. Other fibres terminate in end bulbs, while many
others are distributed to the hair follicles and the sweat
glands. In the deeper or subcutaneous region fibres are
found which end in Pacinian bodies.

The **glands** of the skin are chiefly of two kinds: the
sudoriparous or sweat glands, distributed in varying
abundance over the whole surface of the body; and the
sebaceous glands, found chiefly at the bases of the hairs.
In certain localities specialized glands are found which may
be regarded as modifications of one or the other of these
two types.

The **sudoriparous glands** are situated in the reticular
stratum and the outer portion of the adipose layer: they
are spheroidal bodies, from one-half of a millimetre to
two millimetres in diameter, each consisting of a small
tube coiled into a ball: from this the tube proceeds (as the
duct of the gland) with slight deviations from a direct

course through the corium, at whose surface it becomes continuous with a closely coiled spiral opening through the epidermis; the whole structure consisting, therefore, of a tubular depression of the surface, whose basement membrane is a continuation of that which forms elsewhere the outer limit of the corium, while its lining epithelium is a direct continuation of the epidermis.

The glandular (or the greater part of the coiled portion) is considerably larger than the rest of the tube, being sixty or seventy micra in diameter: immediately upon the basement membrane in this region is found a simple layer of elongated elements resembling smooth muscular fibres in appearance, and commonly regarded as such; their long axes are parallel with the direction of the tube. Upon this, and surrounding the lumen of the tube, is a layer of columnar glandular cells about fifteen micra in diameter, which are frequently pigmented. The duct is from twenty to thirty micra in diameter: it consists of a basement membrane upon which is seated a stratified epithelium of a very simple order: immediately next the membrane is a layer usually two cells thick, composed of polyhedral cells; while upon this is a single layer of flattened cells often designated the cuticle. On reaching the epidermis the basement membrane is, of course, continued into that of the surrounding region: the deeper portion of the epithelium becomes continuous with the stratum Malpighii: the cuticular layer lines the passage through the epidermis, which is from the first over twice the diameter of the lumen of the duct, and in the stratum corneum flares to form a trumpet-shaped opening at the surface.

Each sweat gland has a network of capillaries which penetrates the coil, following the interstitial connective tissue; and is also provided with a proper nerve supply. The number of the glands varies in different portions of the body from four or five hundred to the square inch (or more than a millimetre apart) upon the back of the neck and trunk to five or six time as many (or less than half a millimetre apart) on the palm of the hand or the sole of the foot. Those of the armpits are quite large, as are those at the root of the penis in the male and on the labia majores in the female. About the anus are found glands identical with the sweat glands in structure, but still larger than those just mentioned: they are sometimes distinguished as the circumanal glands. In the larger sudoriparous glands (and sometimes in the smaller) branching of the coiled portion occasionally occurs: and in some cases the duct is bifurcated, either before or after it enters the epidermis. The **ceruminous glands** of the external auditory meatus of the ear resemble the sweat glands in their structure, though differing from them to a marked degree in the nature of their secretion.

The **sebaceous glands** occur, as has been stated, at the roots of the hairs: they are also found upon the labia minores and the prepuce and occasionally in other hairless localities. Each consists of a short duct usually opening into the outer portion of a hair follicle and connecting internally with the adjacent secreting portion: this commonly consists of from three or four to eighteen or twenty saccules, or, rarely, a single spheroidal sac. The basement membrane, which is continuous with that of the corium,

supports a compound layer of polyhedral cells, which in
the glandular portion are constantly being renewed by the
division of the cells next the basement membrane. The
cells thus cast off undergo a sort of fatty degeneration,
the interior of. the gland thus becoming filled with a semi-
fluid mass consisting of oil droplets, fragments of broken-
down cells, etc., which is discharged upon the surface
through the duct. The **Meibomian glands** of the eyelid
are to be regarded as enlarged and modified sebaceous
glands. The mammary glands are also to be regarded as
originally derived from the modification of bodies essentially similar to sebaceous glands and are properly to be
considered as tegumentary organs. On account of their
size and complexity, however, their structure will be better understood after the discussion of the general structure
of glands in a subsequent chapter. They will, therefore,
be discussed later in connection with the female reproductive organs.

Both sudoriparous and sebaceous glands are formed by
solid ingrowths of the epidermis which in the case of the
former penetrate the corium, the cavity being formed in
the centre of the club-shaped mass and later communicating with the surface, while the inner end becomes coiled
into the shape characteristic of the adult structure. In a
similar manner the mass of cells which is the precursor of
the sebaceous gland becomes divided into lobules, the fatty
transformation of the central cells already described beginning in them and advancing along the axis of the pedi-
cle, which thus becomes converted into the duct.

While the **hairs** are practically outgrowths from the
skin they are in reality derived from the modification of
ingrowths resembling in some respects those which give
rise to the sebaceous and sudoriparous glands. A hair is
a mass of epidermal cells which cohere together to form a
cylindrical or more or less flattened rod: these cells are
formed upon and around a **papilla** situated at the bottom
of a depression of the corium known as a **hair-follicle,**
whose sides are lined with a specially modified layer of epidermal cells, continuing the Malpighian layer of the epidermis and the stratum' lucidum to the region around the
papilla at its base. The portion of the hair which projects beyond the surface is termed the **shaft:** that which is
contained in the follicle is known as the **root:** the epider-
mic portion of the follicle surrounding the root is commonly called the **root-sheath.**

The fully formed hair, whether within or without the
follicle, has on its surface a layer of imbricated scales, with
their free edges directed toward the outer end of the hair;
this layer is known as the **cuticle:** on the form and arrangement of the free margin of the cuticular cells depends
the pattern of the superficial markings seen on the hairs of
man and of different animals. Immediately beneath the
cuticle lies the **cortical substance** of the hair: this is composed of greatly elongated fusiform cells in which traces
of the nucleus are still visible, though the body of the cell
has largely undergone horny transformation: these cells
are so closely united that their limits are not ordinarily
distinguishable, the cortex appearing to be made up of
elongated fibres; it is to the color of the cortical cells that

the color of the hair is largely due. The cortical substance in some hairs extends clear to the centre: in others it surrounds (as its name implies) an axial mass of cuboidal or polyhedral cells, the **medulla**: the bodies of these cells usually contain small air-vesicles, rendering the medulla white by reflected light. Such air spaces may also be present in the elongated cells of the cortex.

The formation of the hair at the papilla will be more clearly understood if its description is preceded by that of the follicle with whose epidermis it is continuous: this, like the structure of which it is a modification, consists of a fibrous or dermal portion and an epithelial or epidermal portion. The dermal portion may be resolved into three layers: of these the **outer fibrous layer** consists of bundles of white fibrous tissue running chiefly in the direction of the follicle and containing numerous connective-tissue corpuscles; it resembles the corium in most respects, save in the absence of elastic fibres: the **middle layer**, sometimes called the muscular layer, consists of connective tissue whose fibres run transversely, and of transversely-disposed elongated cells with rod-shaped nuclei which much resemble smooth muscular fibres and have been described as such, but which may perhaps be regarded as modified connective-tissue corpuscles; this layer extends from the bottom of the follicle to the point where the sebaceous gland opens, beyond which it is wanting: the inner or **hyaline layer**, or glassy membrane, as it is sometimes termed, is thin, homogeneous and transparent; it corresponds to the basement membrane of the corium.

The epithelial portion of the follicle, while it is structurally free from the hair within, is generally brought away
with the latter if it be pulled out of the skin during life:
it is therefore commonly termed the root-sheath of the
hair.   It consists of two layers, comparable in a general
way to the mucous and the horny layers of the epidermis.
The outer of these, or **outer root-sheath**, is much thicker
than the inner, being a layer several cells deep; of these
the cells next the hyaline layer (or basement membrane)
are columnar, like those at the base of the stratum Malpighii; while those within are polyhedral prickle cells like
the majority of those in the stratum just mentioned.

The inner layer of the epithelium, or **inner root-sheath,**
may be divided into three strata: the outer of these,
known as **Henle's layer,** is a single layer of flattened cells
of a horny appearance, in which nuclei are not distinguishable: within this is **Huxley's layer,** composed of polyhedral cells with small nuclei, the layer being two or three
cells deep: this is lined on the inside by the **cuticle of the
root-sheath** a single layer of flattened cells which are imbricated in a manner similar to that of the cells of the
cuticle of the hair, but in the opposite direction, the free
edges of the cells in question being directed toward the
bottom of the follicle; as a consequence, their edges interlock with those of the cells on the surface of the hair. The
name of inner root-sheath is sometimes restricted to the
part which comprises Henle's and Huxley's layers, the
two becoming confluent at base of the follicle; the cuticle
of the root-sheath is in such cases described as third layer
thereof; but this is entirely a question of names: no satis-

factory attempt has been made to identify either of these
layers with the various strata of the skin in the way that
the outer root-sheath can be identified with the stratum
Malpighii.

As we pass to the base of the follicle the outer fibrous
layer is continued around its curving extremity to be continued into the **hair papilla,** a large club-shaped papilla
which projects upward into the lower end of the follicle:
the middle layer becomes thinner and terminates near the
lower end of the follicle: the hyaline layer, now resting on
the fibrous layer, invests the surface of the papilla as a
basement membrane. The whole follicle is slightly enlarged
or bulbous at the base; the thickening being chiefly due to
an enlargement of the epithelial portion of the follicle and
of the lower end of the hair itself. The outer root-sheath
is continued downward with little change, the columnar
cells at its base passing around the curve of the valley
which surrounds the base of the papilla and over the surface of that body, acting here as elsewhere as the genera-
tive layer of the epithelium: the polyhedral cells become
confluent with those which form the layers of Huxley and
of Henle, these latter having previously become merged
into one, and with the cells of the cuticle of the root-sheath.
The valley at the base of the papilla is thus filled with a
mass of newly formed cells, which, as they are rapidly
multiplied, are pushed off as a cylindrical mass, the hair,
from around the papilla. The medulla of the hair is formed
from the cells developed upon the upper end of the papilla
itself, and is in a certain sense the continuation of the col-

umnar layer of the outer root-sheath: the cortex of the hair represents the polyhedral layer and the layers of Huxley and Henle; while the cuticle of the hair corresponds to that of the root-sheath.

The outer fibrous layer of the hair follicles is richly supplied with blood vessels and nerves: some fibres of the latter pass to the outer root-sheath, where they appear to terminate among the epithelial cells in a manner similar to that found in the stratum Malpighii of the epidermis, chiefly in the immediate vicinity of the sebaceous glands. In some of the lower animals large hairs, chiefly about the face, are provided with special forms of nerve terminals: such hairs are termed **tactile hairs.**

The hair follicles are rarely vertical to the surface of the skin, the degree of their obliquity varying in different localities and, in consequence, the position of the hair upon the surface. Many hairs have small bundles of smooth muscular fibres passing from a point on the papillary layer of the corium near the opening of the follicle and on the side toward which the hair is inclined, to be inserted in the outer fibrous layer of the follicle near the bulb. The contraction of these muscles, known as the **arrectores pili,** tends to erect the hairs.

Hairs are formed as solid club-shaped downgrowths of the stratum Malpighii of the epidermis, which meet with specially formed papillae around which the hair-bulb is moulded: the young hair is developed as a conical mass

above the papilla, the solid epithelial plug first formed
undergoing sebaceous degeneration in its centre and thus
permitting the escape of the hair: its lateral portions become the root-sheath, outgrowths therefrom giving rise
to the sebaceous glands. When a hair ceases to grow, the
papilla gradually disappears and the hair finally drops
out of the follicle: this may or may not have been preceded by the formation of a new downgrowth from the
bottom of the follicle and the development of a new pa-
illa, thus giving rise to a replacing hair.

The **nails,** like hairs, are masses of epidermal cells, consisting chiefly of a thickened and otherwise modified extension of the stratum lucidum. Each nail can be regarded
as composed of three portions: the **free margin,** in which
growth has entirely ceased, the **nail-body,** which constitutes its greater portion, but which receives but slight additions to its under surface, and the **nail-root,** which is the
region of greatest increase. The body of the nail is continuous below with a modification of the stratum Malpighii, which rests upon a modified portion of the corium,
called the **nail-bed:** laterally this fibrous layer is folded
upward to form the **lateral nail-grooves,** and posteriorly
upward and forward to form the **posterior nail-groove;**
the lower portion of which is termed the **nail-matrix,** including the whitish curved area at the base of the nail
known as the **lunula.**

The nail-bed and nail-matrix are continuations of the
corium which has become highly vascular and is well sup-

plied with nerves: the papillae upon its surface are simple
and closely crowded together: as far as the outer margin
of the lunula they show no definite arrangement, but
throughout the nail-bed proper are arranged in longitudinal rows, their extremities inclining toward the free end
of the nail: they are so closely set in the rows as to appear
to be confluent in ridges, which are sometimes said to replace them. Below, the nail-bed is connected with the distal extremity of the last phalangeal bone by numerous
strong bands of fibrous tissue: as it passes around the margin of the nail to enter the walls of the nail-grooves, it
assumes the structure commonly characteristic of it.

The stratum Malpighii is by some histologists defined
as a part of the nail-bed: it is a question of names merely,
but it is perhaps better on the whole to divide the two regions by the natural boundary between dermal and epidermal structures. The columnar cells of this stratum
are closely packed together, and multiply rapidly, particularly in the region of the matrix: very few polyhedral
cells are to be seen, the newly formed cells passing over
rapidly into the substance of the nail without an intervening stratum granulosum. Like the stratum lucidum,
the body of the nail consists of flattened horny cells, in
which traces of nuclei can be seen after dissociation. During foetal life the nail is invested by the stratum epitrichium, traces of which overlie its margins at birth.

CHAPTER XIV.

THE MOUTH AND ITS CONTENTS.

The **mouth** is formed by an ingrowth from the surface of the head; its lining is therefore epiblastic in origin and directly continuous with the epidermis: the cavity of the mouth does not at first communicate with that of the pharynx, but the two are connected later by the perforation of their common wall at the fauces. Within the cavity of the mouth are found the jaw arches, covered by the gums and bearing the teeth, and the tongue, which rises from its floor; and in the so-called mucous membrane which lines it throughout are found the labial, buccal, palatal and lingual glands: while other and larger glands more remotely situated discharge their secretion into the cavity of the mouth by means of ducts.

As we pass from the skin upon the outer surface of the lip to the **mucous membrane** which is found upon its inner surface marked changes are to be noted in both the epithelial and the fibrous layer. Near the margin of the lip the hair follicles are wanting, though sebaceous glands are present; the derma becomes thinner and highly vascular; while the epidermis becomes much more transparent, allowing the red color of the blood in the dermal capillaries to shine through. As the transition is made from a surface constantly dry from exposure to the air to one constantly moist, the stratification of the epithelium becomes less distinct: and on the inside of the mouth is found a layer of protoplasmic cells corresponding to the stratum Malpighii passing by insensible gradations into a layer of horny cells, flattened, and with small nuclei, that are constantly being exfoliated, corresponding to the stratum corneum: all trace of the intervening stratum granulosum and stratum lucidum disappearing The cells at the base of the layer are columnar, like those in the corresponding portion of the epidermis, and, like them, are constantly forming new cells to replace those lost from the outer surface.

Underneath the epithelium is the fibrous layer, corresponding to the corium, to which the name of **mucosa** or mucous membrane is sometime restricted: it is in most cases thinner than the corium, but bears upon its surface numerous papillae: below it breaks up into looser connective tissue as a rule, the meshes being occupied by the glands of the mucosa and by fat lobules: to this looser subjacent tissue the names of **submucosa** is applied: in the region of the fauces, the soft palate, and the uvula, adenoid tissue is present in great abundance in the mucous membrane; a feature never found in connection with the corium of the skin.

Where the mucous membrane invests the **hard palate**, and where it passes over the arch of either jaw to form the **gums**, the fibrous layer becomes firm and dense, glands and fat lobules alike being absent or very sparingly present, and the deeper portion becomes directly continuous with the periosteum of the subjacent bone. The papillae and the investing epidermis of the dorsal surface of the tongue undergo special modifications best described in connection with that organ.

The term **mucosa,** or its equivalent, **mucous membrane,** above applied to the lining of the mouth, is also used to designate the lining of all those cavities which communicate directly or indirectly with the outside of the body (with the exception of the abdominal cavity of the female). It thus includes the investment of the nasal passages and the associated sinuses, the respiratory tract, the oral cavity and the alimentary canal, the urinogenital tract, and the middle ear with the Eustachian tube. By some histologists the term is even applied to the lining of the hair follicles and the dermal glands. Excluding the latter, it may in general terms be described as consisting, like the skin, and as seen in the lining of the mouth, of an epithelial and a skeletal portion. The epithelium of the mouth and of the nasal passages is epiblastic in its origin: that of the kidneys and genital glands, and of their proper ducts, mesoblastic: that of all the rest of the surfaces named hypoblastic. It is to the regions lined with hypoblastic epithelium that the term mucosa is chiefly applied.

The epithelium of a mucous surface may be simple and flattened, polyhedral, or columnar; or transitional; or stratified. The skeletal layer consists usually of a basement membrane (sometimes termed a **membrana propria,**) which may be either a homogeneous layer of closely felted fibres, or an endothelioid layer of connective tissue corpuscles; beneath this is the felted fibrous layer termed variously the **corium,** the **stroma,** or the **tunica propria,** or, by some, the mucosa in a limited sense. It consists of a thicker or thinner layer of rather loosely felted bundles of white fibres, rich in blood vessels and in some cases having its surface raised into papillae. Elastic fibres are sometimes present in such great numbers as to form a definite elastic layer, and at other times are almost entirely wanting. Adenoid tissue may be present in varying quantity, sometimes forming definite nodules or clusters of nodules. Smooth muscular fibres may also be present, forming a stratum in the deeper part of the membrane one or more layers in thickness and known as the **muscularis mucosae.** In exceptional cases, such as that above described in connection with the hard palate and the jaw arches, the mucosa is firmly united to the subjacent structures: as a rule the deeper portion passes over into a layer of loose areolar tissue known as the **submucosa**, thus permitting of the free movement of the mucosa on the structures beneath.

Where the epithelium of a mucous surface is columnar a varying proportion of the elements have the form of **goblet cells**, as described in a previous chapter: these are perhaps the simplest form of special secreting organs found in the human body; and are sometimes spoken of as **unicellular glands:** the term gland being often loosely employed to designate any secreting organ and, indeed, some organs that are not at all secretory in function.

Increase of secreting surface is obtained by ingrowths of the epithelium which penetrate the fibrous layer to a greater or less extent, in a manner similar to that described in connection with the skin; like those referred to these ingrowths may be either tubular or saccular. Where such ingrowths are not sub-divided internally they are commonly called **simple tubular** or **simple saccular glands**, as the case may be. A distinction of importance, both structural and physiological, ought, however, to be noted in this connection. Some such simple secreting organs are lined with cells that are similar in form and alike secretory in function throughout their whole extent: while in others the work of secretion is restricted to specialized cells in the epithelium of the deeper portions, that which is found upon the portion next the surface having lost its secretory activity and become modified to form the lining of a conducting tube through which the secretion of the deeper portion is discharged. This distinction can be kept in view if we always apply to the structure which secretes throughout its whole extent the name of a **follicle,** and define a **gland** as a **secreting organ** provided with a **duct.**

Glands, as thus defined, may be either **tubular** or **saccular**, and either **simple** or, by the subdivision of the secreting portion, multiple or **compound:** the secreting portion of a simple gland is called the **fundus:** those of a compound gland are called **acini** or **alveoli.** Where the final divisions of a compound gland are saccular in form, the gland is frequently designated as **acinous** or **racemose;** where they are elongated, the gland is called **compound tubular;** and where both forms of alveolus are present the term **acino-tubular** is applied. By many histologists, however, the distinctions which these terms imply are regarded as of questionable value.

Follicles, simple glands, and the smaller compound glands

rarely penetrate deeper than the submucous or subcutaneous tissue of the membrane from which they are derived. Large compound glands, on the contrary, are usually situated at some distance from the surface where their secretions are discharged. Glands which are thus situated outside of the organs to which their ducts lead are called **extrinsic,** the term **intrinsic** being applied to those contained within the organs in question.

With the exception of the tegumentary glands discussed or mentioned in the preceding chapter, all the glands of the body open upon mucous surfaces; and with the farther exception of the kidneys and the genital glands are derived from the modification of a mucous membrane. The secretions discharged by them have always one or both of two functions: to lubricate and preserve the sur_ face in question, which is a general function, or to act as ferments upon food stuffs taken into the body, which is a special function restricted to the alimentary tract. Two principal types of glands are therefore distinguishable both structurally and functionally. Those of the first sort are known as **mucous glands;** those of the second (from the more watery character of their secretions, but less properly) as **serous glands**. In the case of some compound glands some of the alveoli are of the mucous and some of the serous type: such are designated **mixed glands.**

In **glands of the mucous type** the acini or alveoli are lined with polyhedral cells which do not extend quite to the centre of the cavity, thus leaving a well defined central opening or **lumen**: the free extremity of the cell is transparent and does not stain readily with most reagents, while the protoplasm and the somewhat flattened nucleus which it surrounds are crowded down to the base of the cell; in other words, the general appearance is like that previously described as characteristic of goblet cells, which, it should be recalled, are mucigenous in function. In addition, there may in some cases be seen between the glandular cells and the basement membrane crescent shaped groups of granular cells which stain deeply: these were described independently by two observers, who gave them names associated with their form; they are therefore called the **crescents of Gianuzzi**, or the **demilunes of Heidenhain**: their nature and functions are not yet fully understood: the constituent cells (sometimes solitary) are called **marginal cells.**

In **glands of the serous type** the secreting epithelium consists of polyhedral cells which when at rest extend clear to the centre of the alveolus: a distinct lumen cannot, therefore, be recognized. The whole body of the cell is granular, the substance which is to become the characteristic secretion being thus stored up in the protoplasm: the nucleus is spherical, and situated near the centre of the mass; and the whole cell stains readily. No trace of anything like the crescentic cell-masses above mentioned or of marginal cells is to be found in any serous gland.

The appearances above described are those seen in sections of glands previously hardened by reagents, and taken from organs which had not been actively stimulated immediately before the preparation was made. If a piece of the fresh gland be examined in blood serum the mucigen is seen in the case of the mucigenous cells to be present in the form of very large granules: while the ferment secreting cells are so swollen that not only the lumen of the acinus but also the outlines of the cells are obliterated: otherwise the appearances are much as described above. As the result of prolonged stimulation the mucigenous cells discharge their secretion, and the nuclei approach a central position: while the cells of the serous type become smaller after discharge, and a distinct lumen becomes visible in the acinus. The two types of alveoli, become therefore much more nearly alike, though not so much so as to prevent their distinct recognition.

It is ·in the mouth that we first meet with a mucosa; and its discussion has therefore been deferred until after a description of the lining membrane of that cavity. The glands of the mouth also furnish us with examples of both mucous and serous, and both intrinsic and extrinsic glands.

The intrinsic glands of the mouth are the **labial, buccal, palatal** (including those of the uvula), and **lingual**. These are all racemose glands situated in the submucosa, with the exception of the lingual, which lie between the muscular bundles of the tongue. They are all of the mucous type save those on the posterior part of the tongue, which are serous. Their secretion contributes to form the saliva of the mouth.

The extrinsic glands of the mouth are those commonly referred to under the name of the **salivary glands**: they are designated, from their positions as the **sublingual,** the **submaxillary,** and the **parotid glands**: the first-named are doubtfully to be called extrinsic, on account of their structure and position, since they consist in each instance of a group of small glands opening by several ducts, and are situated just beneath the mucous membrane on either side of the base of the frenum of the tongue. The nature of the salivary glands differs in different mammals: in man the sublingual and submaxillary are mixed glands, and the parotid is of the serous type.

The acini or alveoli of one of the larger or extrinsic **salivary glands** vary greatly in form from flask-shaped sacs to wavy, contorted, or even branched tubules: the basement membrane is reticulated and the epithelium varies with the type of gland in question. Each alveolus leads into a tubule smaller than itself known variously as the **ductule,** the **connecting** or **intermediate tubule** or the **intercalary duct**: the basement membrane of the ductule is continuous with that of the alveolus; its epithelium consists of a single layer of flattened cells. The ductules of a number of alveoli enter a common tube known as an **intralobular duct** or **salivary tube of Pflueger,** the alveoli and ductules with the duct just mentioned together making up a **lobule.** An intralobular duct consists of a basement membrane continuous with that of the ductules, and a single layer of columnar epithelial cells: the latter have spherical nuclei situated near the centre of the cell;

the outer extremity of the cell is finely granular, while that next the basement membrane is vertically striated. The basement membrane of the alveoli, the ductules and the intralobular duct alike rest upon the interstitial connective tissue, which contains rich networks of capillaries surrounding the alveoli.

The intralobular ducts lead into larger conducting tubes known as **interlobular ducts**, around which the lobules are aggregated into **lobes,** their boundaries within the lobe being marked by septa of fibrous tissue. The interlobular lead into the **interlobar ducts,** which finally enter the **chief duct** of the gland. The larger ducts have beneath the basement membrane a definite fibrous layer which in some cases contains smooth muscular fibres: the epithelium is columnar and simple, save in the largest ducts, in which a layer of polyhedral cells lies between the columnar cells and the basement membrane.

The interstitial connective tissue which fills the spaces between the alveoli passes into thin laminae of fibrous tissue which separate the lobules: these **septa** are again continuous with stouter structures of the same nature which lie between the lobes of the gland. The interlobar septa are continuous internally with the stroma of connective tissue which immediately surrounds the proper fibrous tunic of the chief duct and its principal subdivisions, and externally with a membranous layer which surrounds the whole gland and is known as its **capsule.** The place where the chief duct leaves the interior of the gland is known as the **hilum:** the nerves, arteries, veins, and lymphatics of the gland also enter or leave at this point,

the capsule here becoming continuous with the stroma of
fibrous tissue (above mentioned as surrounding the duct
and its branches) in which they lie, and in which occasional small nervous ganglia may be found.

The **saliva** contained in the mouth is a mixture of the
secretions of the various intrinsic and extrinsic glands.
With the nature and properties of the fluid itself we are not
here concerned: it constantly contains, however, certain
histological elements which may with propriety be mentioned in this connection. The most abundant of these
are the squamous cells which are constantly being exfoliated from the surface of the stratified epithelium of the
mouth: they occur singly and in patches; and when found
in the latter form the overlapping of their bevelled edges
can be plainly seen: the nuclei are small and flattened, and
stain readily. Less numerous, but quite abundant in the
saliva from the back part of the mouth are the so-called
salivary corpuscles, spheroidal bodies but little larger
than colorless blood corpuscles, each containing one or
two spheroidal nuclei and numerous minute granules which
exhibit a constant dancing movement within the interior
of the cell. The salivary corpuscles are really modified
leucocytes that have escaped into the oral cavity from the
tonsils or similar adjacent structures, and have become
swollen by the imbibition of the watery saliva; the granules of the protoplasm being suspended in the imbibed
fluid and exhibiting Brownian motion in consequence.

The mouth contains certain organs by means of which

the food is masticated and prepared for swallowing by the thorough admixture of the saliva. These are the teeth and the tongue. A **tooth** is a vertically elongated mass of the tissue mentioned in a previous chapter by the name of **dentine,** having a solid free portion, the **crown,** projecting above the gum and covered with a layer of **enamel:** and a hollow portion, the **fang,** imbedded in a socket in the jaw known as an **alveolus;** the outer surface of the fang is covered with a layer of **cementum,** and the internal cavity occupied by the **pulp.** The slightly constricted region where the crown and the fang meet and the tooth pierces the gum is called the **neck** of the tooth.

The **pulp** which fills the central cavity of the tooth is a mass of connective tissue approaching in character more nearly to the gelatinous tissue of the embryo than any other structure in the adult human body, its soft and sparingly fibrillated matrix containing numerous branching corpuscles whose processes are connected. It contains a rich network of bloodvessels, and a small bundle of nerve fibres, entering its substance through the minute canal at the tip of the fang. Toward its surface the corpuscles rapidly increase in number and in size, forming a superficial layer of crowded cells much resembling a columnar epithelium : these are the **odontoblasts;** they are in all probability directly associated with the formation of dentine.

Like osseous tissue, **dentine** is characterized by a lamellated and calcified matrix traversed by canaliculi: it differs essentially from the substance of bone, however, in the mode of calcification and the definiteness of the lamel-

lae, and particularly in that the characteristic corpuscles
associated with its formation, the odontoblasts, do not,
like the osteoblasts of bone tissue, become imbedded between the lamellae, but remain situated upon the inner
surface of the mass, their long and sparingly branched
processes, the **fibres of Tomes,** occupying the canals, called
dentinal tubules, which traverse the dentinal lamellae.
Calcification takes place, at least in the outer portion of
the mass, by the deposition of globular nodules of lime
salts; these at first do not occupy the whole of the matrix,
leaving numerous irregularly stellate spaces, the **interglobular spaces,** the outer layer in which they abound
being known as the **granular layer.** The interglobular
spaces communicate outwardly with the surface of the
dentine and inwardly with the dentinal tubules: corpuscles have been described as contained in them and communicating with the extremities of the fibres of Tomes.
Deeper the calcification becomes more nearly continuous;
at certain intervals it is, however, incomplete, irregular
lines in a general way parallel to the surface being seen:
these, known as the **incremental lines of Salter,** may be
regarded as marking the boundaries of the lamellae. The
dentinal tubules are lined by a condensation of the fibres
of the matrix sometimes described as a distinct membrane;
they have a wavy or spiral course across the lamellae and
give off branches occasionally at acute angles which pur-
sue a similar course; their waviness gives rise to a striated
appearance when a tooth is seen in longitudinal section
with the naked eye, or a low power of the microscope:
the alternating dark bands seen are known as the **lines of**

Schreger. The general direction of the tubules is in the fang at right angles to the lamellae; in the crown they pass obliquely upward.

The **enamel** consists of a layer of elongated calcified prisms, usually hexagonal in cross section, set in a general way vertical to the surface of the dentine of the crown of the tooth; they are alternately lighter and darker throughout their extent, giving to the enamel as a whole a banded appearance. In addition, occasional dark brown lines may be seen crossing the enamel columns parallel-wise to the surface; these, known as the **stripes of Retzius,** may indicate lines of growth. Here and there vertical spaces may be seen slightly separating the enamel columns near the dentine, with which the interglobular spaces of the latter possibly communicate. At the time of irruption the surface of the enamel is covered with a thin cuticular layer known as **Nasmyth's membrane;** it is rapidly worn away as soon as the teeth are put to active use.

The **cementum,** cement-substance, or **crusta petrosa,** as it is variously termed, investing the surface of the fang, is composed of tissue essentially similar in structure to that of dense bone: the lamellae are few and irregularly arranged, the lacunae varying in size and form; the canaliculi of the latter are said to communicate with the interglobular spaces of the adjacent dentine: there are no clearly defined Haversian systems. The layer is formed by the so-called **periodontal membrane,** which is practically continuous outwardly with the periosteum lining the alveolus which contains the tooth.

The enamel is epithelial in its origin, being derived from
the calcification of a layer of cells or formed by their secretion. The dentine and cementum are true skeletal structures. The margin of the foetal jaw early shows a thickening of the stratified epithelium which grows downward
into a groove into the mesoblastic tissue beneath, the
dental groove; the curved rod of epithelial cells thus
formed is known as the **primary enamel germ.** At regular intervals along the under side of this rod further prolif-
erations of the epithelial cells occur, with localized downgrowths, the **special enamel germs**, at first flask-shaped.
At the same time a condensation in the mesoblastic tissue
beneath each gives rise to a conical **dental papilla** which
grows upward to meet the epithelial downgrowth. The latter grows down about the papilla, thus becoming converted
into a cap-like mass, the **enamel organ**, connected with
the epithelial ridge above by a slender stalk: the stalk is
attached to one side of the enamel organ, owing to the
mode of growth of the latter. A condensation of the
mesoblastic tissues surrounding the newly formed enamel
organ and papilla gives rise to a membrane which
soon becomes rich in bloodvessels, and is known as the
dental sac.

The papilla, thus invested by the enamel organ, assumes
the general shape of the crown of the future tooth: the
corpuscles near the surface become more numerous and
larger than those of the interior, and the layer of odontoblasts is formed, after which the deposition of dentine is
begun. Important changes have in the meantime taken
place in the enamel organ: the cells upon the concave sur-

face next the papilla become greatly elongated vertically, and by their activity begin the deposition of the enamel; whether by a process of secretion or by the transformation of their own substance is still a matter of debate; the layer of enamel-forming cells is known as the **inner layer** of the enamel organ: the margin of this layer is continued over the convex surface of the enamel organ by the **outer layer** of cuboidal epithelial cells, which is in turn continuous through the stalk with the epithelium of the jaw, and which lies in contact with the dental sac: the interior of the enamel organ contains a mass of cells, the so-called **middle layer,** which becomes converted into branching corpuscles, their interspaces being filled with a watery fluid. From the side of the enamel organ or from the adjacent stalk a bud arises early in the history of the tooth, and grows downward to form a second enamel organ in the case of those teeth which are succeeded by others.

As the time of irruption approaches, the middle layer of the enamel organ becomes greatly compressed, and the enamel forming cells of the inner layer reduced in depths to flattened scale-like bodies, forming the membrane of Nasmyth. The papilla rapidly elongates, forming the fang, thus pushing the first formed crown upward toward its final position. After the dentinal portion of the fang is fully formed the cementum is deposited upon its outer surface in a manner practically the same as that of the formation of bone elsewhere.

The **tongue** is a mass of striped muscular tissue invested with the mucous membrane of the mouth, which upon its upper surface is specially modified. The muscular fibres which make up its bulk are divided into two symmetrical masses by a median vertical partition of fibrous tissue, the **lingual septum**, which is better developed below than above: right and left of this they lie in interwoven bundles some of which run longitudinally, others transversely, and a third set vertically; their arrangement to form the lingual muscles is a matter anatomical rather than of histological discussion.

The mucous membrane of the lower surface and sides of the tongue does not differ essentially from that of the rest of the mouth. At the margin of the upper or dorsal surface, however, it suddenly becomes modified, the majority of the papillae becoming greatly enlarged to form the projections of the surface visible to the naked eye and known as the **lingual papillae**. Of these there are three kinds: those most abundant, and covering the whole surface of the tongue with a velvety layer, are known as **conical** or **filiform** from their shape; they are tapering upgrowths of the corium, covered with a corresponding layer of epithelium, and are frequently bifid at their tips: scattered here and there among them are spheroidal elevations of the surface about the size of a small pin's head; these are termed **fungiform,** and are invested with a thin layer of stratified epithelium. Near the root of the tongue, and arranged in a V-shaped figure whose apex is directed backward, are eight to twelve large papillae, roughly cylindrical in form, which are sunk in depressions of the surface and surrounded by circular trenches or grooves; these are the

circumvallate papillae: the stratified epithelium of their sides forms a deep layer in which are imbedded the taste-buds or gustatory organs. The description of these structures will be given in the chapter devoted to the organs of special sense.

The papillae all contain capillary networks, the fungiform and circumvallate papillae having a specially abundant blood supply. The mucous membrane of the posterior part of the tongue contains diffuse adenoid tissue with here and there occasional nodules of the same. The lingual glands of the anterior part of the tongue are all of the mucous type: some of those of the posterior part, known as the **glands of Ebner,** are of the serous type, their ducts opening into the grooves which surround the circumvallate papillae. The tongue is formed in part by an upgrowth from the floor of the mouth, in part by a growth forward from the ventral wall of the pharynx: its epithelium is therefore in part epiblastic and in part hypoblastic in origin; but the two regions cannot be distinguished in the adult

CHAPTER XV.

THE ALIMENTARY CANAL.

As we pass from the mouth through the opening of the fauces we enter the tube lined with hypoblastic epithelium of which the mouth is the antechamber, and to which the name of the alimentary canal is strictly applicable. The anterior and posterior pillars of the fauces are folds of the mucous membrane rich in mucous glands and containing some diffuse adenoid tissue: between them on either side of the mouth is situated a **tonsil**: a rounded body of variable size consisting of a mass of fibrous tissue richly infiltrated with adenoid tissue and containing a variable number (from ten to twenty) well-defined nodules of the same structure, like those found in the mucous membrane of the rest of the tongue. The surface of the adenoid mass thus formed is covered with stratified epithelium continuous with that adjacent, which is in places invaded by numbers of leucocytes passing through it into the oral cavity to become the bodies described above as salivary corpuscles. Deep folds or depressions of the surface occur, lined with stratified epithelium: these are termed **crypts**, and receive the secretions of subjacent mucous glands.

The **alimentary canal**, properly so-called, is a tube which, beginning at the pharynx and ending with the rectum, varies in size and form in different portions. Its structure varies correspondingly in detail, as will be stated in the descriptions of the various regions. In general plan it is a tube having essentially a double wall whose divisions are connected by a stratum of areolar tissue, and invested throughout a portion of its extent by the serous membrane which lines the abdominal cavity.

The inner or glandular layer, the **mucosa,** is a mucous membrane whose epithelium varies greatly in different parts: the fibrous portion is in all the various regions more or less infiltrated with adenoid tissue which is in places quite scanty in amount or wanting altogether, in places so abundant as to form conspicuous nodular masses. Smooth muscular fibres are also present in all the regions, usually in such quantity as to form a well-defined **muscularis mucosae,** which may consist of two or even three distinct layers.

The stratum of areolar tissue into which the loose and folded mucosa passes, the **submucosa,** varies in extent in different regions, but always permits of free movement of the folds of the mucosa. It is characterized throughout the greater portion of its extent by the presence of a network of nonmedullated nervous fibres with small ganglia at the nodes, the **plexus of Meissner:** and contains the larger blood and lymph vessels from which those of the mucosa are derived.

Beyond the submucosa, which is continuous with its internal skeletal framework, is the outer principal layer of the tube, the **musculosa,** or muscular layer. The anterior portion is composed of striped fibres; the remainder (and greater portion) of smooth fibres. Except in the pharynx, where definite muscles are found, the bundles of fibres, whether smooth or striped, are arranged in two continuous strata, the inner of which is composed of circular and the outer of longitudinal fibres: between them is a thin layer of connective tissue containing blood vessels and a second and larger nervous network known as the **plexus of Auerbach**.

In the neck and thorax the connective tissue adjacent forms a more or less well defined membranous layer just without the musculosa sometimes described as a separate layer of the canal under the name of the **fibrosa**; while the term **serosa** is commonly applied to the peritoneal fold which rests upon the musculosa throughout the abdominal cavity.

The **pharnyx** is the first region of the alimentary canal. Its upper or respiratory portion is lined with ciliated columnar epithelium; while that of the lower and larger portion is stratified squamous: the latter is throughout a large part of its extent more or less infiltrated with leucocytes. The fibrous membrane contains a large amount of adenoid tissue, which in the upper and posterior part of the pharynx is collected into a large patch containing adenoid nodules and crypt-like depressions, to which the name of the **pharyngeal tonsil** has been given: glands of the mucous type are abundant. Beneath the submucosa is found a dense layer of fibrous tissue, the **pharyngeal aponeurosis,** separating the mucous membrane from the muscular coat. Without are the well defined **pharyngeal muscles,** composed of striped muscular fibres. The structure of the pharynx may perhaps be best understood if we regard the musculosa proper of the alimentary tract as entirely deficient, the aponeurosis representing the fibrosa, and the constrictor muscles as external additions thereto.

The **oesophagus**, like the lower part of the pharynx, is lined with stratified squamous epithelium. The mucous membrane, which is thrown into longitudinal folds, bears numerous small papillae upon the surface, covered by a homogeneous basement membrane. Adenoid tissue is very sparingly present in the mucosa, but occasional solitary nodules may be found. The muscularis mucosae is wanting in the uppermost part of the oesophagus: lower it at first appears in the form of scattered longitudinal bundles, which near the lower end of the oesophagus become so numerous as to form a continuous layer of longitudinal fibres.

The **submucosa,** which is well developed, contains the intrinsic glands of the oesophagus, which are of the mucous type; their ducts traverse the mucosa: blood vessels and lymphatics are numerous: the plexus of Meissner is wanting or very scantily developed throughout the greater part of the oesophagus.

The **musculosa** consists throughout the upper third of a transverse and a longitudinal layer of striped muscular fibres; lower down the layers are made up wholly of smooth fibres. The plexus of Auerbach is present, though its ganglia are smaller and less numerous than those of the rest of the alimentary canal. There is a more or less well defined fibrosa just without the external muscular layer.

The oesophagus opens into the **stomach**, a saccular dilation of the alimentary canal whose wall is greatly modified, chiefly in the **mucosa**: this layer is much thickened, owing to the development of long tubular glands which open upon its surface, are imbedded in it, and make up the greater part of its substance; it is thrown into numerous folds, or **rugae**, and is covered with a simple layer of columnar epithelium containing numerous goblet cells, the transition between which and the stratified epithelium of the oesophagus is abrupt: its surface is dotted all over with the minute openings of the gastric glands previously mentioned.

The **gastric glands** are distinguished by their structure into two kinds, designated according to their position as **cardiac** (also called **peptic**), occupying the region extending from the oesophageal opening to beyond the middle of the stomach, and the **pyloric**, which occupy the lower third. The former have short ducts lined with columnar epithelium similar to that of the surface of the stomach: the glandular portion is a long wavy or slightly coiled tubule, whose end is sometimes strongly bent; from two to four such secreting tubules commonly opening into a single duct: lining the tubule throughout its whole extent is a single layer of polyhedral cells, while scattered along the tubule here and there between the layer of cells just mentioned and the basement membrane are larger spheroidal or ovoidal cells. The cells first mentioned are called the **chief** or from their position the **central cells,** or, from their probable ferment-secreting function, the **peptic cells:** the others are known as the **accessory** or **parietal cells;** they have also been called the **oxyntic cells,** and are supposed to produce the acid of the gastric juice.

The pyloric glands differ from the cardiac in the proportions of their parts; the ducts being much longer and the tubules shorter: in the form of the secreting tubules, which are much more contorted; and in the absence of parietal cells. Between the region of well defined cardiac and pyloric glands is a narrow **intermediate zone** of transition in which the ducts become longer and the tubules shorter than in the cardiac region, while the parietal cells become less numerous and finally are wanting altogether.

The interstices between the gastric glands are filled with interstitial areolar tissue containing a rich network of capillaries and lymphatics and a small amount of adenoid tissue, with occasional slender bundles of smooth muscular fibres. About the bases of the secreting tubules the adenoid tissue is more abundant, and here and there forms small nodules with ill-defined boundaries. Below, there is a well-developed muscularis mucosae, consisting of two distinct layers, an inner transverse and an outer longitudinal layer: the inner is to some extent subdivided into two laminae, the fibres of one running somewhat obliquely to those of the other; from its inmost surface the slender bundles above mentioned as running up between the gastric glands are given off.

The **submucosa** of the stomach is a rather thick layer of areolar tissue which extends upwards into the folds of the mucosa which form the rugae. It contains the larger bloodvessels and lymphatics connected with the vascular supply of the mucosa, and a well developed plexus of Meissner situated nearer to the mucosa than to the muscular coat.

The **musculosa** is thicker in the stomach than in any other portion of the alimentary canal, the increase in thickness being chiefly in the inner or transverse coat. This, in addition to being greatly increased in volume is in the cardiac portion more or less clearly divided into two layers lying obliquely to each other and to the longitudinal coat. The plexus of Auerbach is conspicuously seen between the two coats. On the outer surface is the **serosa**, a thin fibrous membrane covered by endothelium, and adhering closely to the longitudinal muscular coat except along a narrow strip at the attachment of the mesogaster and the omentum.

As we pass from the stomach to the **duodenum** the mucosa becomes much thinner, the glands present in the upper portion of the latter being situated in the submucosa; a narrow zone of transition is seen between the pyloric glands of the stomach and the duodenal glands, or glands of Brunner. The **mucosa,** which is thrown into folds, the valvulae conniventes, is covered with finger-like or leaf-like projections, the **villi,** between which are tubular depressions, the **intestinal follicles** or **crypts of Lieberkuhn**: the surface of the villi and the lining of the crypts are alike covered with simple columnar epithelium con-

taining large numbers of goblet cells, the columnar cells
having vertically striated caps or borders. Beneath the
epithelium of both crypts and villi is an endothelial basement membrane, sometimes called the **subepithelial endothelium of De Bove.**   .

The centre of a villus contains a lymphatic with a
widely opened extremity: around it are scattered bundles
of smooth muscular fibres proceeding from the muscularis
mucosae below: near the surface is a network of capillaries formed from the breaking up of a small artery at the
base of the villus and uniting to form a vein near its tip:
the interstices are filled with loose adenoid tissue. Between
the crypts at the base of the villi is a rich network of capillaries and lymphatics and imbedded, like those of the villi,
in adenoid tissue, which becomes quite dense about the
bases of the crypts; here and there well defined nodules
are found.

Immediately beneath the layer of adenoid tissue upon
which the crypts abut is situated the muscularis mucosae,
which is not so thick as in the stomach, but presents two
distinct layers, an inner transverse and an outer longitudinal.

The **submucosa** of the upper portion of the duodenum is
quite thick: in addition to the blood vessels and lymphatics
and the plexus of Meissner, it contains the **glands of
Brunner** already mentioned. These are compound tubular glands, whose short alveoli, lined with polyhedral cells
surrounding a distinct lumen, open into slender branching
ducts which pass to the centre of the gland, there to open

into the central duct: this, which is lined with short columnar cells, penetrates the mucosa to open between the villi. The glandular epithelium resembles that of the pyloric glands, between which and the glands in question there is a distinct transition at the pylorus: the epithelial cells contain fine granules similar to those found in the cells of glands of the serous type, which these glands resemble save in the non-essential feature of the larger size of the lumen.

The transverse layer of the musculosa, as it reaches the lower end of the stomach, becomes suddenly thickened to form a muscular ring, the **pylorus** or **gastro-duodenal valve**: passing into the intestine it becomes as quickly reduced in thickness, forming a layer much thinner than that found in the gastric wall: a similar diminution is found in the thickness of the longitudinal layer. The musculosa of the small intestine does not otherwise differ from that of the stomach, like which it is invested by a serosa formed from the peritoneum.

In the lower portion of the duodenum the glands of Brunner are wanting, and the submucosa becomes greatly reduced, the canal assuming the structural character common to the greater portion of the **small intestine,** whose division into the regions recognized by the anatomist is not marked by any histological characters. The only noteworthy feature in this respect is the **specialization** seen in the quantity and disposition **of the adenoid tissue** of the mucosa, which attains its highest development in the lower portion of the small intestine, the ileum.

Throughout the whole of the small intestine adenoid tissue is found in the villi and between the crypts, as in the
upper portions of the duodenum just described : and as in
that structure, there is a continuous layer thereof between
the crypts and the muscularis mucosae. This, however, becomes much thinner and less dense, while here and there
appear well defined **nodules** commonly known as **solitary
follicles** or (by the use of a term equally unfortunate) **solitary glands**: it should of course be understood that these
bodies have functionally nothing in common with either
glands or follicles as defined on a preceding page : they are
broadly pear-shaped bodies whose bases usually extend
into the submucosa, which is locally thickened in connection therewith, and whose apices when large project as
rounded eminences upon the surface of the mucosa covered
with a single layer of columnar epithelium, both crypts
and villi being wanting at that point: in structure each is
a dense mass of adenoid tissue in the main uniformly disposed, but with a slight diminution in density toward the
centre of the mass, containing a capillary network and surrounded by numerous lymphatics: the nodules are surrounded by the fibrous tissue of the mucosa, which is here
quite dense, but they have no definite capsule.

The solitary nodules of adenoid tissue give place to a
great extent in the lower portion of the intestine to clusters of similar bodies sometimes called **agminated glands,**
but better known as **Peyer's patches.** Where these occur
the mucosa and submucosa are both thicker than in other
portions of the intestine : the nodules are closely crowded
together and in the centre of the mass become confluent:

the conical apices of most of them project upon the surface of the mucosa: the epithelium of the projections (as is the case to some extent with that of the solitary nodules just described) becomes infiltrated with leucocytes as in the case of that investing the tonsils, often to such an extent as to be no longer clearly distinguishable, large numbers of leucocytes thus constantly entering the intestinal cavity.

At the margin of the folds which constitute the ileocolic valve the villi cease abruptly; and as we pass the margin in question we come upon the wall of the **large intestine**, including its greater portion, the colon, with its anterior prolongation, the caecum (ending in the vermiform appendix), and its posterior continuation, the rectum.

The **mucosa** of the **colon**, like that of the rest of the large intestine, is devoid of villi. It is crowded throughout its whole extent with tubular follicles resembling the crypts of the small intestine, but larger and showing ordinarily a larger proportion of goblet cells. The interstitial adenoid tissue is scanty, but large solitary nodules occasionally occur. The muscularis mucosae does not differ materially from that of the small intestine.

The **submucosa** is a layer of moderate thickness, containing the larger vessels and the plexus of Meissner. The **musculosa** is well developed: the inner or circular layer is thickened uniformly, while the outer layer is chiefly gathered into three longitudinal bands. The outer surface is covered for a portion of its extent with a **serosa**.

The **caecum** is essentially similar in structure to the colon. The **appendix vermiformis** is a rudimentary structure, variable in size and development: its mucosa contains numerous follicles, like those of the colon, and also solitary nodules of adenoid tissue: its musculosa consists of two layers, the outer or longitudinal differing from that of the colon in being of uniform thickness and thicker than the inner or circular layer.

As the alimentary canal is continued downward to form the **rectum**, its structure undergoes important changes in each of the fundamental layers, in connection with the modification which takes place in the function of the canal: these will as before, be described in connection with the layers successively.

The **mucosa** of the rectum, like that of the colon, is devoid of villi: it is much thicker than that of the colon, increasing in thickness downwards to the anal opening. In addition to temporary foldings, chiefly longitudinal, which exist when the tube is empty, there are present three or four transverse folds, semilunar in shape, known as the **valves of Houston**, and near the terminus several longitudinal folds, the **columns of Morgagni**; their relations to subjacent structures will be mentioned later. The epithelium of the greater portions of the rectum is columnar, like that of the small and the large intestines: just above the anus there is in man a transition from simple columnar to stratified squamous epithelium continuous with that of the surface of the body: this transition takes place upon the columns of Morgagni higher up than within the depressions which lie between them. Follicles or crypts like those of the colon abound in the mucosa of the greater portions of the rectum: they disappear at the level of the commencement of the columns of Morgagni. In addition to a small amount of diffuse adenoid tissue, occasional solitary nodules are found: they are less numerous relatively than in the colon.

The fibrous structure of the mucosa of the rectum is denser than that of the colon, particularly toward the lower extremity, where there is a marked increase in the proportion of elastic fibres and a diminution in the amount of adenoid tissue, the fibrous layer finally passing without abrupt transition into the corium of the skin. The **muscularis mucosae** is gradually increased in extent, and consists chiefly of longitudinal fibres: at the lower portion these are gathered together in several stout bundles, which, with the overlying folds of the mucosa, form the columns of **Morgagni** above described : these are continued into tendinous bundles which are inserted into the skin in the immediate vicinity of the anal opening.

The **submucosa** of the rectum becomes considerably thickened toward its lower extremity, in connection with the increased mobility of the mucosa, and is in addition reinforced by the presence of numerous elastic fibres : it contains, moreover, a certain amount of adenoid tissue, in this respect differing markedly from the same layer in other portions of the canal. A distinct plexus of Meissner is present.

The **musculosa** of the rectum undergoes marked increase in thickness from above downward in both the circular and the longitudinal layer. In the former, there is in
addition to the gradual increase in volume a tendency to
divide into bundles of fibres of gradually increasing size:
at its lowermost limit a number of these bundles are aggregated together to form the **internal sphincter** muscle
which surrounds the anal opening: distinct thickenings of
the circular layer also underlie the valves of Houston. The
longitudinal layer, as it passes downward, also undergoes
subdivision, the resultant bundles of fibres diverging more
and more from the circular layer, and being roughly divisible into three strata; an inner, the bundles of which are
interwoven with those that form the internal sphincter; a
middle, whose bundles terminate along the boundary of
the external sphincter (an extrinsic muscular structure
composed of striped fibres); and an outer stratum whose
bundles unite with the levatores ani and the recto-coccy-
geal muscles.

The extrinsic glands connected with the alimentary canal
are but two in number, the pancreas and the liver, both of
them opening into the duodenal portion of the small intestine, of which they are outgrowths. The **pancreas** is in
most respects similar in structure to a large salivary gland
of the serous type; its connective tissue framework as well
as its blood and lymph vascular supply being the same,
and the form of the alveoli similar, as well as their arrange-

ment in lobules and lobes. The epithelium of the alveoli resembles that of a serous gland in its granular appearance,
and in the position of the nucleus; the lumen is also very
small. Certain differences are, however, to be noted. The
alveoli are as a rule longer and more generally tubular in
form: the epithelial cells are taller and narrower, as a rule,
approaching to the columnar form: there have been described under the name of **centroacinar cells,** certain spindle-shaped elements occupying the lumen near the union
of the alveolus with the intermediate tubule; their origin
and function are alike uncertain. The intermediate tubules,
with their characteristic flattened epithelium, are more
extensive than in the salivary glands, replacing in great
measure the intralobular ducts. The pancreas is also
characterized by the presence between the alveoli of peculiar groups or clusters of **intertubular cells** which form
nodular masses situated in the intralobular connective
tissue or the interlobular septa; each mass having a definite tuft of capillaries: these bodies whose nature is not
clearly understood, are by some regarded as newly formed
or embryonic alveoli.

The **liver** is at once the largest and the most highly
modified of all the digestive glands. While it must be regarded as formed by the modification of a primarily tubular structure, the multiplication of the glandular epithelial cells has obliterated the lumen of the tubule in each
instance, the secretion formed by the epithelium being discharged through the agency of intercellular channels which
are unlike anything found in any other portion of the
human body. This extreme modification in the arrange-

ment of the glandular epithelium is also accompanied by
an equally great modification in the arrangement of the
blood vessels, the capillary network apparently penetrating the epithelial masses; thus forming what at first sight
seems to be an exception to the general rule that epithelia
are devoid of blood vessels. The structure of the liver will
therefore best be understood if we begin by examining the
blood supply and the framework of skeletal tissue with
which it is associated.

The surface of the liver is invested by a thin **capsule** of
fibrous tissue, surmounted by the peritoneal endothelium.
From this capsule thin lamellae penetrate the surface of
the gland as **interlobular septa**: in the transverse fissure
it becomes continuous at the **porta** (which corresponds to
the hilum of an ordinary gland) with an important ingrowth of connective tissue, the so-called **capsule of Glis-
son,** which forms by far the larger part of the interlobular
tissue, becoming continuous with the ingrowths from the
capsule. The porta is the place of entrance for the nerve-
supply of the liver, and for the large **portal vein** and the
smaller **hepatic artery**; as well as of exit for the **bile
duct** and for the principal **lymphatic trunk**: these four
vessels and their ramifications, together with strands of
nerve-fibres, are usually found in close proximity, surrounded by a certain amount of connective tissue, the
whole constituting what is known as a **portal canal.**
The blood leaves the liver by means of the **hepatic veins,**
whose branches are formed by the union of the **sublobular
veins,** the origin of which will be presently described: the
sublobular and hepatic veins are not accompanied in their

course by branches of other vessels, a character by means of which they can be readily distinguished from the branches of the portal vein.

The **lobules** of the liver, sometimes (but less correctly) termed the *acini* or alveoli, are masses of hepatic cells, penetrated by capillary networks, polyhedral in form as the result of pressure, and in man something over a millimetre in diameter. Those at the surface of the liver have the axis vertical or nearly so to that surface: but there is no such regularity of position in the case of those more deeply situated. The interlobular septa are in a few mammals (of which the pig is one) quite stout and conspicuous: in man they are far less developed; and the boundaries of the lobules are in consequence not always readily determined. From certain points on the septa a very scanty framework of connective tissue traverses the interior of the lobule, accompanying the capillaries: its presence even in small quantity is a fact of importance in forming a clear conception of the essential structure of the liver.

The branches of the portal vein end in veinlets situated in the septa and hence known as **interlobular veins**, each lobule having several surrounding it: these give off capillaries (in a manner similar to arteries elsewhere) which form a network whose meshes converge toward the axis of the lobule: this is occupied by a single vessel, the **central** or **intralobular vein**, into which the capillaries empty. The intralobular vein passes at the base of the lobule into a **sublobular vein**, the hepatic veins being as has been stated, formed by the confluence of the sublobular veins.

The branches of the hepatic artery accompany those of the portal vein to the lobules, where they divide into inter-lobular arterioles situated in the septa, together with the interlobular veins: like the latter, they terminate in capil-laries; these are, however, situated chiefly in the connec-tive tissue of the septa and the walls of the veins, a por-tion of them only penetrating the lobules for a sufficient distance to form channels of communication with the capillary network of the lobule, as a means of discharge of the blood from the arterioles. The portal vein and its final branches, the interlobular veins, are therefore to be regarded as the channels by means of which the blood is submitted to the glandular action of the hepatic cells; the hepatic artery being probably in large part, if not alto-gether the channel of nutrition for the skeletal framework and vascular mechanism of the organ.

The network of capillaries with its accompanying scanty connective tissue forming, according to some observers, perivascular lymph channels, is interwoven with another and somewhat coarser network, that of the strands of liver cells or hepatic cords. These are polyhedral epithelial cells about twenty micra in average diameter, with granu-lar protoplasm and central spherical nuclei, which perform the complex secretory function of the liver. The size of these cords are such that nearly every cell is brought at some point in contact with a capillary; the surface of the cell being in many cases slightly excavated along the line of contact.

Between the hepatic cells is found a network of minute passages usually not more than a micron in diameter, the so-called **bile capillaries** or, as they are otherwise termed the **bile canaliculi**: these are so situated that they traverse the surface of the cell along a side or an angle opposite that in contact with the blood capillaries, never passing between the latter and the cells. At the surface of the lobule the bile canaliculi communicate with small **interlobular bile ducts** whose flattened epithelium is continuous with the hepatic cells, and whose basement membrane is resolved into the scanty connective tissue of the lobule. These unite as they pass to the portal canals to form the smaller bile ducts situated therein; the epithelium becoming columnar and the basement membrane stouter. As the smaller bile ducts come together to form the larger trunks found in the more central region of the liver the cells of the columnar epithelium become taller and are seen to rest upon a distinct membrana propria: while the fibrous layer beneath is reinforced by smooth muscular fibres: the largest ducts have in addition small mucous glands in their walls. The structure of the duct leading to the intestine, and of the gall bladder as well, is essentially the same as that of the largest ducts within the liver.

If we recall the structure of a gland of the serous type it will be remembered that the epithelial cells which line the alveoli are so large as to reduce the lumen to a very slender and often imperceptible channel between their apices: the bile canaliculi may with propriety be compared to such channels. The alveoli of ordinary glands may be either spheroidal or elongated and tubular, the interstices between them containing a variable amount of connective tissue which supports a capillary network, whose meshes conform to the structure and arrangement of the alveoli. In the lower vertebrates the liver consists of tubular alveoli essentially similar to those of other glands. If we conceive of such tubules in the mammalian liver as becoming elongated and branched, and finally anastomosing, we shall understand the structure and arrangement of the hepatic cords. According to this view the anastomoses formed obliterate the boundaries of the alveoli; but we may regard as belonging to one such division all the bile canaliculi with their surrounding epithelium which lie nearest to and presumably discharge into one of the interlobular bile ducts: the beginnings of the ducts, with their flattened epithelium, corresponding in position and structure with intermediate tubules or intercalary ducts of other glands: the lobule is thus seen to be in fact as well as in name a lobule; that is, a collection of acini or alveoli; and the application of either of the latter titles to it is misleading. The most peculiar features in the structure of the liver are those connected with the blood supply.

CHAPTER XVI.

THE RESPIRATORY APPARATUS.

In the description of the alimentary canal which has
just been given, the conspicuous layers, whose modifications in the various regions have been discussed, are those
most readily demonstrable by the knife of the anatomist :
they are those, moreover, which naturally result from the
double function of the greater part of the canal, the inner
layer, or mucosa, being in substance a glandular layer by
means of which the nutritious portion of the food is
brought into a condition suitable for the accompanying
process of absorption; the outer layer, or musculosa,
being in effect a mechanism by which the food mass is
caused to traverse this glandular surface; while the submucosa makes possible the necessary movements of the
mucosa and the musculosa on each other. We should,
however, bear in mind the continuity of the submucosa
with the fibrous structure of the mucosa on the one hand,
and on the other with the interstitial connective tissue of
the musculosa, and through it with the fibrosa or serosa :
considering this, and bearing in mind the embryonic origin
of the tissues present, it will be quite clear that another
very natural division of the wall of the alimentary canal,
based on histological rather than anatomical characters,
would recognize two primary layers of tissues (differing

vastly in extent), the epithelial layer, hypoblastic in its
origin; and the musculo-skeletal (and vascular) layer derived from the mesoblast, extending from the basement
membrane, just beneath the epithelium within, to the
fibrosa (or, in the intestine, to the fibrous basis of the
serosa) without: the presence of this fibrous layer as one
of the constituents of the alimentary canal should be
clearly borne in mind.

It is by an outgrowth from the pharyngeal region of
the alimentary canal that the respiratory apparatus is
formed, and its structure throughout its whole extent is derived from such modifications of a musculo-skeletal layer
lined with epithelium as are required for the proper performance of the functions of the various regions. These
are in the trachea and bronchial tubes such as will serve to
maintain the patency of these channels for the passage of
the air and in some measure to regulate the same: and in
the airsacs such as will permit the freest interchange of
gases between the air which they contain and the capillaries within their walls.

The **trachea** is plainly seen to be composed of three
well-defined layers, similarly disposed to those which appear as the chief factors of the alimentary canal. The
inner of these is the **mucosa**; it differs in some important
respects from the layer of that name in the region just
mentioned. It is lined with what is usually termed a
stratified columnar ciliated epithelium: the layer is but
a few cells deep; those at the surface are cylindrical or
prismatic, with tapering bases, their free extremities in
most cases beset with numerous fine cilia, whose united

action causes constant currents toward the mouth; here and there goblet cells occur; beneath the superficial cells, or interspersed between their bases, are elongated and more or less spindle-shaped cells, which are probably destined to replace them; still lower are pyriform and spheroidal cells, many of which multiply rapidly by cell-division, replacing the older superficial cells as occasion requires, resembling thus in function the deepest portion of the stratified squamous epithelium of the oesophagus.

The epithelium of the trachea rests upon a **basement membrane,** a homogeneous, elastic layer of considerable thickness, perforated by occasional fine canals. This is followed abruptly by a **fibrous layer,** whose bundles are irregularly and somewhat loosely disposed and are intermingled with a considerable number of elastic fibres: there is a well but not greatly developed network of small blood vessels, accompanied by nerves and lymphatics, and a considerable amount of adenoid infiltration. Beneath the fibrous layer, and continuous therewith, is the **elastic layer,** composed of a dense network of elastic fibres, most of which are longitudinally disposed, which form the inner boundary of the mucosa: it is best developed in the dorsal portion of the trachea.

The **submucosa,** like that of the alimentary canal, is a layer of areolar tissue serving to unite the mucosa with the denser layer beyond it. In addition to the larger blood vessels from which the blood supply of the mucosa is derived, and their associated lymphatics, it contains numerous small **glands** of the mucous type, whose long ducts, lined with cuboidal epithelium, traverse the mucosa to open on its surface. These glands are, as a rule, most abundant in those portions of the submucosa which are situated opposite the intervals between the successive fibrous rings of the trachea. Here and there in the submucosa an occasional adenoid nodule may be found, frequently associated with a gland or its duct.

No single word will readily indicate the complex structure of the outer layer of the trachea: if we for convenience make use of the term **fibrosa** in referring to it, we shall designate it by its most constant though not most conspicuous factor. It is in effect a tube of dense fibrous membrane, rich in elastic fibres, reinforced at regular intervals by the incomplete rings of cartilage which are the most prominent components of the trachea, and bearing on the inner surface of its dorsal portion the rudiments of a muscular layer. The rings are composed of hyaline cartilage which is far less brittle than usual: they are imbedded in the fibrous layer, which is continuous with the perichondrium of each, and situated rather toward its inner than its outer limit. The muscular layer occupies an elongated rectangular area nearly identical with that bounded by the lines connecting the dorsal ends of the incomplete cartilaginous rings, but somewhat more extensive laterally: to the band of muscular tissue thus formed, the name of the **tracheal muscle** is given. It consists entirely of smooth muscular fibres; the great majority of these are arranged in transverse bundles disposed in groups which are separated by occasional transverse septa of fibrous tissue; external to these are a few thin and short longitudinal bundles forming an imperfectly defined layer:

the muscular layer is invested outwardly by the fibrous
layer which it in some measure replaces, the latter being
here much thinner than in other portions of the trachea.
The bundles of the transverse layer are inserted between the
rings upon the fibrous layer; those opposite the rings are
in man inserted on their inner surface; in some of the lower
mammals they are inserted on the ends and in others on
the outer surface of the rings. The longitudinal bundles
are inserted upon the ends of the rings and upon the fibrous layer. It is a noteworthy fact that in the dorsal portions of the trachea some of the mucous glands are situated external to the muscular layer, their ducts penetrating it as well as the other layers of tissue beneath the epithelium.

The **bronchi,** or right and left divisions of the trachea,
while they differ from it anatomically, particularly in the
form of the cartilages, resemble it in all essential respects
as regards their histological structure. As they enter the
lungs they branch and subdivide repeatedly, their immediate continuations and their subdivisions, with the exception of the smallest, being known as the **bronchial
tubes,** or, as they are sometimes termed, the **intrapulmon-
ary bronchi:** by the time they are reduced to a diameter
of one millimetre, or thereabouts, they are known from
their size and structure as **bronchioles;** the smallest of
these are never less than half a millimetre in diameter.

The largest **bronchial tubes** are essentially like the
extra-pulmonary bronchi in structure: as we pass toward
the bronchioles, their structure undergoes marked though

gradual modification: that of a tube of medium size may be described as follows. The **epithelium,** like that of the trachea and larger bronchi, is stratified, columnar and ciliated, and rests on a homogeneous basement membrane. Beneath this is a **fibrous layer** containing a moderate amount of adenoid tissue, and numerous elastic fibres no longer forming a continuous layer, but gathered into strands, which form the bases of the folds into which the inner portion of the mucosa is frequently thrown: external to the adeno-fibrous layer is a well-defined **muscularis mucosae,** composed of smooth fibres transversely disposed, this layer, which is of variable thickness from point to point along the tube, is sometimes designated the **bronchial muscle.**

External to the muscular layer is the **submucosa,** composed here, as in the trachea and bronchi, of areolar tissue containing mucous glands and plexuses of blood and lymph vessels. The outermost layer, that which we have for convenience termed the **fibrosa,** is the one which first shows signs of reduction: the fibrous membrane is by no means as dense, nor as clearly defined, either from the submucosa within or from the adjacent tissues without, as in the trachea; the cartilaginous rings of the latter are represented by plates of cartilage of varying size and shape; and there is nothing present in this layer that may be regarded as corresponding structurally to the tracheal muscle. Occasional lobules of fat may be seen, and the mucous glands not unfrequently penetrate into this layer, which thus tends to approach in its structure that of the submucosa.

As we pass toward the smaller bronchial tubes the epithelium becomes gradually reduced to a single layer of columnar ciliated cells resting upon a basement membrane. The fibrous layer beneath becomes much thinner; the relative amount of adenoid tissue becomes less and less, and elastic fibres become far less numerous, though they do not altogether disappear. The muscular layer, on the other hand, for a time at least, increases in relative thickness, becoming one of the most conspicuous features of small tubes. The submucosa and fibrosa become blended into one layer of loose fibrous tissue rich in lymphatics and containing blood vessels, the mucous glands disappearing from the former, together with the cartilaginous plates (and the nodules which succeed them) from the latter.

Within the **bronchioles** (otherwise designated the **terminal bronchi**) still further reductions of structure take place: the epithelium changes from columnar to cubical, loses its cilia, and later becomes more or less flattened, forming a single layer of polyhedral granular cells upon the basement membrane. The latter rests on a thin layer of fibrous tissue with longitudinal elastic fibres: the muscular layer is reduced to scattered bundles and later to isolated fibres, without finally disappearing altogether; while the submucosa and the fibrosa become blended with each other and with the fibrous layer of the mucosa.

The bronchioles lead into larger pyramidal or irregular shaped spaces, the **infundibula,** into each of which open by wide apertures a large number of the spheroidal or

polyhedral air-sacs or **alveoli** which are the ultimate
chambers of the lung: each infundibulum with its associated alveoli making up one of the component **lobules**
of that organ. The principal change in passing from the
bronchiole to the infundibulum is found in the epithelial
layer: the low granular cells with which the distal extremity of the bronchiole is lined are found in the infundibulum
in patches which become smaller and less numerous as we
proceed to its farther extremity: between them are found
larger and thinner transparent cells which form a simple
squamous epithelium to which the distinctive title of **respiratory epithelium** is applied. The basement membrane,
the longitudinal network of elastic fibres, and the layer
of scattered muscular fibres are continued without essential change. The fibrous portion of the wall of the bronchiole is represented by scattered branched connective tissue corpuscles situated in the interstices between the elastic and muscular fibres.

The **alveoli** are in the main continuations of the walls of
the infundibulum. The epithelium consists almost entirely
of the large flat cells above mentioned, the smaller granular cells being scattered sparingly among them either singly
or in groups of two or three: between the cells, and in particular at the angle where three or four meet, are occasional
stomata, minute openings which communicate with the
lymph spaces below. About the mouth of each alveolus
there is an annular bundle of elastic fibres from which is
given off a network, which, together with a small amount
of fibrous tissue and a few connective tissue corpuscles,
forms the wall of the alveolus and the support of the epi-

thelium and capillary network: from the form and disposition of the alveoli it results that a single layer thus
formed does duty in great measure for two adjacent alveoli.
The **capillary network** contained in the interalveolar
septum thus formed is exceedingly dense; and its loops pass
from side to side of the septum in a serpentine course, thus
bringing the blood contained within them as near as possible to the air in each of the alveoli.

The **lymph spaces** in the alveolar walls communicate
with the lymphatics situated in the connective tissue
septa which lie between the lobules. Each lobule, as thus
bounded, is irregularly pyramidal in form, its apex being
situated at the bronchiole, toward which the limiting
septa tend: peripherally, the interlobular septa are continuous with the denser layers of fibrous tissue which
form the investment of the lobes: the latter, in their turn,
being continuous with the pleura which forms the serous
investment of the surface of the whole lung.

The **larynx** is a special modification of the proximal end
of the trachea: it differs in its histological structure from
the latter in the following particulars. Its epithelium is
in most portions stratified, columnar and ciliated, as in
the trachea: that of the true vocal cords and of the epiglottis and a portion of the intervening surface is stratified squamous: in that covering the under surface of the
epiglottis numerous taste buds are imbedded. The mucosa
differs from that of the trachea, chiefly in the greater
amount of adenoid tissue, in places resembling the pharynx in this respect: the elastic fibres which run throughout its extent are greatly increased in number in the vocal

cords, where they constitute the chief part of the mucosa. The principal cartilages are hyaline, like those of the trachea: those of Santorini, of Wrisberg and of Luschka are reticular, as is the epiglottis. The proper muscles of the larynx are composed of striped fibres, resembling in this respect the musculosa of the upper portion of the oesophagus.

The course and arrangement of the **blood vessels** of the lung are subjects rather for anatomical than histological discussion. It should be borne in mind, however, in studying sections of that organ, that the lung has a double blood supply: the branches of the **pulmonary artery** accompanying the bronchial tubes to their common destination in the lobules of the lung; while the smaller **bronchial arteries,** derived from the aorta, are distributed to the walls of the air passages themselves and to the surrounding structures, as vessels of nutrient supply. A similar relation subsists between the **bronchial veins,** which empty into the ascending vena cava, and the branches of the **pulmonary veins.**

The **lymphatics** of the lung are divisible in a somewhat similar manner into two groups, those associated with the bronchial tubes, and those associated with the air sacs. The former, or **bronchial lymphatics,** have their origin in the lymph spaces of the bronchial mucosa, forming a plexus in the submucosa which empties into larger trunks leading to the root of the lung. The second group is compound of two sets, the **deep** or **vascular lymphatics,** having their origin in the connective tissue of most of the lobules of the lung, and the **superficial** or, as they are also termed, the **subpleural lymphatics,** which arise in the vicinity of the lobules near the surface and enter into the plexus which underlies the pleura: the latter communicates with the thoracic cavity by means of occasional stomata. The deep and superficial plexus alike empty into trunks which lead to the root of the lung, there to enter, in company with those from the bronchial lymphatics, into the **bronchial lymph nodes.**

It is a common custom to speak of the lung as essentially similar in structure to a gland, the alveoli being compared with the similarly named divisions of the latter, the infundibula and bronchioles to the ductules and intralobular ducts, and the bronchial tubes to the larger ducts. Comparative anatomy shows that the primary condition of the lung is that of a large sac-like outgrowth of the alimentary canal, its unquestionable homologue in the fish-like vertebrates being the swim-bladder. In some of these forms it undergoes more or less subdivision by foldings of its inner surface, and assumes something of a respiratory function; its connection with the alimentary canal remaining simple and entirely membranous: in the lower air-breathing vertebrates the area of respiratory surface is increased to a limited extent only by peripheral sacculation, while a rudimentary trachea and larynx appears: it is only in the higher reptiles, the birds, and the mammals, that the lung assumes the compact and spongy structure due to compound sacculation and associated with the presence of a well developed system of bronchial tubes. Its origin is therefore seen to be the reverse of that of a large gland, which is developed by an increase in the number of originally small alveoli.

While the respiratory tract is to be regarded as a sub-divided saccular outgrowth of the alimentary canal, we must not expect to be able to recognize essential similari-ty of detailed structure in organs differing markedly in function: a general comparison may, however, aid in understanding and remembering the character and ar-rangement of the tissues present. The epithelium is, of course, continuous throughout. The mucous membrane of the alimentary canal, with its varying amount of ade-noid tissue and its subjacent muscular layer, is represented in the air-passages by the adeno-fibro-elastic layer, to which in the bronchial tubes a muscularis mucosae is added. The submucosa of one is continued into that of the other without essential modification. The outer layer of the larger air-passages may best be regarded as corresponding to the fibrosa of the pharynx and oesophagus, reinforced by the tracheal and bronchial cartilages. The musculosa of the alimentary canal is represented only by the muscu-lar area of the trachea with its well developed transverse and rudimentary longitudinal layer.

CHAPTER XVII.

THE URINARY ORGANS.

The urinary organs include the glandular kidneys; their ducts, the ureters; the bladder, to which the latter lead; and the urethra, through which the contents of the bladder are discharged. The urethra of the female is strictly a urinary tract, and is throughout its entire extent a portion of the ventral outgrowth of the alimentary canal from a part of which the bladder is eventually formed: it will therefore be described in this connection. The prostatic portion of the male urethra includes all that part of the tract homologous with the urethra of the female; its remaining portion is a common channel for the urine and the seminal fluid, and is formed in connection with structures accessory to reproduction: the description of its whole extent will therefore be deferred until it can be taken up in connection with them in a subsequent chapter.

If a **kidney** be cut through from the convex surface to the hilum, either longitudinally or transversely, the surface of the cut section shows to the unaided eye certain features of importance to the study of its histology. The solid portion is seen to be curved around a central cavity, the **pelvis** of the kidney; into this project a number (varying with the direction of the section) of conical **papillae**, each of which is surrounded by a cup-like extension of the pelvic cavity termed a **calyx**. The solid portion itself can be readily seen to be divided with considerable sharpness into an outer region, the **cortex**, forming about one third of the depth, and an inner, the **medulla**; the latter can be again distinctly subdivided into the **boundary layer** next the cortex, and forming about one fourth of the entire depth, and the **papillary portion,** which includes the remainder. The papillary portion is distinctly but uniformly striated: in the boundary layer radial tracts, termed **medullary rays,** similar in appearance to those of the papillary substance alternate with tracts characterized by increased transparency.

The entire mass of radiating tracts extending from the apex of a papilla to the meeting of the cortex and the boundary layer constitutes what is known as a **pyramid of Malpighi**; there are rarely less than ten or more than twenty of them in a single human kidney: in some of the lower mammals they are numerous, as in man; in others there is a single papillary ridge which projects into the pelvis along its length, or a single large central papilla: kidneys such as the latter are termed umpyramidal. If the pyramids of Malpighi be conceived of as extended across the cortex, the kidney would be divided into a corresponding number of parts commonly called lobules: if, however, we are to compare these glands with others on the basis of the arrangement of their ducts, these regions will with greater propriety be termed **lobes**: in the human kidney they are not structurally separated, but in that of some mammals, as the otter, each is invested by a capsule of its own, and is but slightly attached to its fellows: a condition characteristic of the human kidney during foetal life. In all multipyramidal kidneys, like that of man, the cortical portion of each lobe extends around and beyond the base of the pyramid toward the pelvis of the kidney : these interpyramidal masses, composed in part of portions of the cortical substance of the adjacent lobules, and in part of blood vessels whose origin and course will be presently described, are known as the **columns of Bertin.**

The medullary rays seen in the boundary layer are less readily traceable in the cortex : they are, however, continued into that layer, growing smaller as they approach the outer surface, and disappearing altogether before reaching it. These prolongations of the medullary rays are known as the **pyramids of Ferrein:** their bases, as will be readily understood, rest on the bases of the pyramids of Malpighi, each of which gives rise to a large number of them. The intervening portions of the cortex make up what is known as the **labyrinth:** a definite portion thereof, not marked off by any visible separation, is structurally continuous with each pyramid of Ferrein. Each pyramid with its associated portion of the labyrinth, and its continuation by means of the medullary ray to the apex of the pyramid of Malpighi, may, as we shall see, be with propriety regarded as corresponding in extent to a **lobule** of other glands. Scattered throughout the labyrinth are small rounded bodies, the **Malpighian corpuscles,** barely visible to the naked eye. Beyond the cortex may be readily seen the tough fibrous capsule which invests the whole gland.

The kidney, like the liver, is characterized by the posses-

sion of a specially modified blood supply, the most important features of which are visible to the naked eye or with
a low power of the microscope. If a section be made as
above indicated through a well injected organ the following facts may be noted. The **renal artery**, entering at the
hilum, divides into four or five branches which traverse the
lining of the pelvis, their subdivisions entering the columns
of Bertin. As they pass along the latter they give off twigs
to the cortical substance present, and on reaching the level
of the bases of the pyramids of Malpighi give rise to an
arched plexus through whose meshes the pyramids of Ferrein pass to the cortex. From the **arcuate arteries** of this
region branches are given off which run to the surface of
the kidney, known as the **radiate** or (from their position
between the cortical areas surrounding the pyramids of
Ferrein) the **interlobular arteries**: from these are given
off twigs which supply the cortical substance in a manner
presently to be described: their extremities terminate in
the capillary network of the capsule of the kidney. From
the arcuate arteries also arise near the point of origin of the
interlobular branches slender vessels, the **arteriae rectae**,
which supply the medullary portion, passing directly
toward the apex of the papilla.

There may also be seen in the cortical portion of the kidney, and receiving twigs which proceed from its substance,
radiate or **interlobular veins** situated in close proximity
to the arteries of that name. They arise just beneath the
the capsule by small vessels having a stellate arrangement
(the **stellules of Verheyen**), and traverse the cortex to enter
into **arcuate veins** in the main similarly disposed to the

arteries of the boundary region; they resemble the latter in receiving **vena rectae** from the medullary portion, and unite to form trunks which pass by way of the columns of Bertin to traverse the lining of the pelvis and come together at or near the hilum to form the **renal vein.** The vasa recta pass toward the papillae in groups which alternate in the boundary layer with the medullary rays, forming the tracts of greater transparency above mentioned.

The kidney is a **compound tubular gland** made up of lobes and lobules which, as we have seen, are not sharply defined from each other by fibrous septa. Its component tubules differ in a marked degree from those of any other gland in the definiteness of their course; the variation in size and in the character of the epithelium of the different regions of each; and in the peculiar relations which they sustain to their blood supply. With the exception of the blood-vessels and their accompanying lymphatics and nerves, together with the small amount of connective tissue in which these are imbedded, and of a very scanty interstitial tissue, the whole substance of the kidney, both cortical and medullary, is made up of these tubules.

Each **uriniferous tubule** is made up throughout its whole extent of a homogeneous basement membrane lined with a single layer of epithelium. It has its origin in a Malpighian body, the extremity of the tubule being there expanded to form a thin walled sac whose distal portion is inverted into the proximal, forming a **capsule of Bowman:** into the cavity of invagination thus formed is thrust a spheroidal mass of capillaries termed a **glomerulus:** its relations will be described later. The capsule is lined throughout with a simple squamous epithelium; and with the accompanying glomerulus makes up the Malpighian body. The tubule leaves the capsule at a point opposite the inversion for the glomerulus by a constricted **neck** lined with small cuboidal cells, expanding at once to form the **proximal convoluted tubule**: this is lined with cuboidal cells, whose outlines are irregular and interlock in such manner as to render it almost impossible to distinguish the boundaries of adjacent cells when seen in longitudinal section; the portion of the cell next the basement membrane is vertically striated, and is, under the influence of certain reagents, separable into rod-like processes: the free surface bears cilia-like processes projecting into the lumen. The tubule passes by sweeping curves toward the pyramid of Ferrein: as it enters it it turns toward the boundary layer in wavy curves as the **spiral tubule of Schachowa**: its structure is essentially like that of the convoluted tubule, the epithelium being distinctly "rodded."

Entering the boundary layer, the tubule suddenly becomes much smaller and nearly straight in its course toward the papilla: it is here known as the **descending tubule of Henle**: this is the smallest portion of the whole tubule; its epithelium has the form of plate-like cells, whose nuclei cause a central thickening which projects into the conspicuous lumen, making its course apparently irregular. Shortly after reaching the papillary portion the tubule bends upward, forming **Henle's loop**, the diameter increases slightly, and the epithelium assumes the form of low polyhedral cells, greatly reducing the lumen. From the loop the **ascending tubule of Henle** passes upward through the boundary layer without essential modification, save a slight increase in diameter, as a straight or slightly wavy tubule.

The ascending tubule enters the cortex and travels for a longer or shorter distance in the pyramid of Ferrein, finally bending abruptly to enter the labyrinth as an **irregular tubule,** whose size and the position of whose lumen varies greatly in accordance with irregular variations in the height of the striated epithelium, the lumen remaining minute throughout its whole extent: it pursues a zigzag course toward the Malpighian body whence the tubule arose. In the vicinity of the latter it resumes a nearly uniform diameter and exchanges its angular course for regular curves: it is now known as the **distal convoluted tubule:** in size, form, and structure, it is identical with the proximal. Like the latter, its course tends toward the pyramid of Ferrein; as it proceeds, it narrows into a **junctional tubule,** or, as it is also called, an **arched collecting tubule,** with low cuboidal epithelium and a relatively large lumen. It passes toward the axis of the pyramid to enter a **straight collecting tubule** somewhat larger, but otherwise similar in structure, which may also receive other junctional tubules from point to point as it passes toward the boundary layer.

The straight collecting tubule passes along the medullary ray through the boundary layer unchanged in form or size: after it enters the papillary portion it joins at an acute angle with similar tubules, the resultant tubules becoming larger with each union : the largest (and terminal) tubes so formed, known as the **ducts of Bellini,** are lined with a simple columnar epithelium : their openings, visible to the naked eye, are scattered over the apices of the papillae.

The **interlobular arteries,** as has been previously stated, pass from the boundary to end in the capillaries of the cortex. Along their course through the cortex they give off on every side short arterioles which go to the neighboring Malpighian bodies : these, which are known as the **vasa afferentia,** on reaching these bodies give rise to the spheroidal masses of capillaries called **glomeruli,** the arteriole entering the glomerulus at a point opposite the neck of the capsule : the capillaries unite again to form small vessels, the **vasa efferentia,** which leave the glomeruli where the afferent vessels enter.

The efferent vessels, it will be seen, sustain to the glomeruli the relation of veins : they do not, however, proceed to the interlobular veins, but divide shortly after leaving the glomeruli to give rise to the **cortical network of capillaries,** which is made up throughout the labyrinth of short irregular meshes interwoven with the convoluted and irregular tubules, throughout the pyramids of Ferrein of long meshes running in the direction of the pyramid. The capillary network is not distinguishable by any known means into areas corresponding to the Malpighian bodies, the individual tubules, the lobules, or even the lobes of the kidney, being apparently continuous throughout the cortex of the whole kidney.

Here and there in the capillary network the radicles of small veins are formed, chiefly in the vicinity of the **interlobular veins**, to which they lead. The veins in question, arising from the stellate veins just beneath the capsule, gather in their course the cortical veinlets and pass to the arcuate veins of the plexus lying in the boundary. They thus form the channels of return for the blood which leaves the interlobular arteries by way of the vasa afferentia, flowing, as should be noted, through two sets of capillaries, those of the glomeruli and those of the cortex, as well as though the intervening vasa efferentia.

The **vasa recta,** as has been stated, leave the arteries and veins of the boundary in close proximity to the interlobular vessels, in some cases springing as branches from their bases. They pass in the transparent striae of the boundary layer to the papilla, the arteries being resolved along the way into the **medullary capillaries,** which form a network with greatly elongated meshes running in the direction of the tubules between which they chiefly lie: in the tip of the papilla the meshes are much shorter, forming a denser network about the ducts of Bellini. The capillaries are gathered up into the radicles of the veins, which lie in close proximity to the arteries. Occasionally the small vessels which arise from that portion of the cortical network of capillaries nearest the boundary layer, instead of entering the interlobular veins, pass downward into the medulla: these, which are known as **false vasa recta,** sooner or later divide again into capillaries which enter into the medullary network. The vasa recta divide the medulla into regions corresponding to the lobules.

The **capsule** of the kidney is a thin but tough fibrous membrane, somewhat lamellated in structure, especially toward the outer surface, and having in its deep portion a scanty plexus of smooth muscular fibres. At the hilum the capsule is continuous with the fibrosa of the ureter. Its surface is invested with a small amount of adventitious areolar tissue which in some cases contains numerous fat lobules, and by which it is connected ventrally with the fibrous layer of the peritoneum, and dorsally with the fascia of the adjacent muscles. A plexus of **lymphatics** has been described in the capsule whose channels connect with spaces in the superficial portion of the cortex: it unites at the hilum with lymphatics which accompany the blood vessels along and from the medullary rays.

The lining which invests the **pelvis** of the kidney and its prolongations, the **calyces,** may be regarded as the continuation of the inner coats of the ureter, and may, like the wall of the latter, be regarded as consisting of an epithelial and a musculo-skeletal layer. The epithelium is of the **transitional** type, which is found only upon urinary surfaces: its peculiarities will be discussed in connection with the bladder, in whose structure they are most readily demonstrable. Immediately beneath is a thin but dense fibrous membrane which, with the epithelium, makes up the mucosa of the pelvis. This constitutes the sole investment of the papillae, the two layers of the mucosa being continuous with the epithelium and basement membrane respectively of the uriniferous tubules at the mouths of the ducts of Bellini. As we pass along the surface of the calyx, however, the mucosa begins to be separated from the surface of the columns of Bertin by scattered muscular fibres, which increase in quantity as we approach the margin of the fold which surrounds the tip of the papilla: here a well defined muscular ringis formed, which has been compared to the transverse layer of the musculosa; it may be regarded as in a certain sense a sphincter of the papilla. The general surface of the pelvis is lined by the mucosa and the muscular layer, there being present between the two an inconspicuous submucosa of areolar tissue, which also contains a small amount of adenoid tissue and scattered mucous glands.

The **ureter,** as has been stated, consists of an epithelial and a musculo-skeletal layer; the components of the latter being so distributed as to form with the epithelium a mucosa, a submucosa, a musculosa, and a fibrosa. The mucosa is a continuation of that just described as lining the pelvis of the kidney, with which it agrees in structure in every essential respect: it is relatively greater in quantity, being thrown, like that of the oesophagus, into longitudinal folds. The submucosa is correspondingly increased in quantity : as in the bladder, it contains a small amount of diffuse adenoid tissue, scattered nodules having also been described, as have occasional mucous glands. The musculosa shows throughout the whole length of the ureter two well defined layers, an inner longitudinal and an outer transverse; and along the lower portion traces of a third layer, external to the circular, are found, in the form of scattered longitudinal bundles. Each of the principal layers contains a comparatively large amount of interstitial connective tissue as compared with the muscular

layers of the alimentary canal: outwardly this is continuous with a well defined but not very dense fibrosa.

The **bladder** resembles the ureter in the essential histological structure of its wall. The epithelium which lines
the mucosa resembles that of the pelvis of the kidney and
of the ureter, as has been stated: the term **transitional**
generally applied to it, has, as it is generally defined, little
meaning; it serves, however, to connote certain characteristics pertaining to epithelium found only on urinary
surfaces, and distinguishing it from stratified squamous
epithelium, with which it is often compared. The most
marked pecularity of transitional epithelium is the power
possessed by all of its cells (and not the deeper layers only,
as is the case with stratified squamous epithelium) of
changing and regaining its form in connection with the
stretching or relaxation of the membrane beneath: this
can be most readily seen by comparing sections of the collapsed bladder and of one distended by a hardening fluid:
both conditions will therefore be described.

In the former case the transitional epithelium is seen to
consist of three distinct forms of cells arranged in what
may perhaps be termed as many layers, though, as will be
seen, the boundaries which separate them are not strongly
marked. The surface is invested with a single layer of cells
which are almost spheroidal or cuboidal above (though
usually not as deep as they are wide), the upper surface
being either flattish or slightly convex when the mucosa is
neither stretched nor pressed together, and almost hemi-
spherical when the bladder is strongly contracted: the
lower surface of the superficial cells is sculptured by con-

cavities which fit closely the surfaces of the cells of the next layer: two nuclei are sometimes seen in a single cell, and it is quite possible that the superficial cells still retain the power of cell-division. The next layer consists of large pear-shaped cells, their rounded ends fitting into the excavations upon the under surfaces of the superficial cells, and their smaller tapering extremities extending to the basement membrane beneath. The spaces between the large ends of the pyriform cells and the membrane are occupied by the constituents of the third layer, which is one or two cells deep, and consists of smaller closely packed spheroidal or polyhedral cells.

If the bladder be distended, the whole epithelial layer becomes much thinner: the superficial cells become flattened, approaching squamous cells in form; they never, however, loose their characteristic sculpturing: the pyriform cells become greatly shortened: and the spheroidal cells correspondingly flattened. A knowledge of the appearance of the elements in both these conditions is highly important on account of the frequent appearance of bits of epithelium in morbid urine. It should be remembered also, that these changes doubtless take place regularly with the daily periodic changes in the state of the bladder. It is well, too, to note the fact that the cells which compose transitional epithelium are always closely in contact, whatever changes of form they may undergo: and also that this epithelium has a remarkable power of resisting diffusion into the blood vessels beneath of the soluble constituents of the urine.

The **mucous membrane** of the bladder is much thicker

and firmer than that of the ureter, and the transition to the submucosa is much more abrupt. As in the ureter, a muscularis mucosae is wanting, while a small amount of adenoid tissue is present. The **submucosa** contains numerous elastic fibres, and occasional mucous glands, particularly toward the base.

The **musculosa** consists throughout of smooth muscular fibres whose bundles are in a general way (and particularly at the equator of the bladder) arranged in three layers, a middle circular and an outer and inner longitudinal: the bundles are, however, quite irregularly disposed, and there is much interstitial connective tissue: it is, therefore, not always easy to recognize the layers. At the base of the bladder the circular layer is increased in quantity to form the internal sphincter. The interstitial connective tissue is continuous externally with a rather loosely woven **fibrosa,** which is over a part of the outer surface of the bladder invested with serous endothelium.

The **female urethra** continues the wall of the bladder to the mucous membrane of the vestibule. In its course the epithelium passes from transitional to stratified squamous, the remainder of the mucosa undergoing no important change. The submucosa contains numerous elastic fibres and near the bladder a number of mucous glands: its deeper layer is highly vascular. The musculosa consists of an inner circular and an outer longitudinal layer of smooth fibres, and there is no well defined fibrosa.

CHAPTER XVIII.

THE MALE REPRODUCTIVE ORGANS.

The male and female reproductive bodies, or **gonads,** are in their origin, mode of development, and primary position essentially similar bodies. The reproductive elements produced in them differ so widely, however, in their activities as to call for widely differing mechanisms for their discharge: mechanisms which are nevertheless derived from the modification of intimately allied structures, whose development is always closely associated with that of the urinary apparatus. For this reason the description of the early stages of the latter will be deferred until the completion of this and the subsequent chapter.

The male gonads, or **testes,** each with its associated **epididymis,** are in the adult human subject suspended by the **spermatic cords** in saccular folds of the skin confluent below the penis and forming the **scrotum.** The latter is formed, as will be more fully described later, by the pushing down of the abdominal wall on either side of the base of the penis: its structure therefore conforms to that of the body wall, subject, however, to special modifications in each of its constituent layers.

The **skin** of the scrotum is thin, corrugated, rich in brownish pigment and in sebaceous glands, and has scat-

tered over its surface flattened curling hairs with conspicuous bulbs. The superficial fascia of the groin and adjacent
parts is continued into the scrotum to form a characteristic tunic, the **dartos,** which is best developed in the fore-
part of the scrotum: it is quite highly vascular, and contains numerous smooth muscular fibres; it is consequently
of a reddish-brown color. The right and left dartos tunics
unite in the mid-plane to form a partition, the **septum
scroti.** The corrugation of the skin of the scrotum is
caused by the constriction of the muscular fibres of the
dartos.

Immediately internal to the dartos is a denser and firmer
fibrous layer, thin and transparent, known as the **spermatic fascia:** it is derived from the tendon of the external
oblique muscle of the abdominal wall. Within and closely
associated with the spermatic fascia is a layer of areolar
tissue which contains numerous bundles of striped muscular fibres variously disposed, constituting a more or less
continuous muscular layer termed the **cremaster,** and
formed from an extension of the external oblique muscle.
Still farther within, the fascia transversalis is continued
in the scrotal wall by the **infundibuliform fascia,** a fibrous layer separated from the preceding by loose areolar
tissue and immediately underlying the parietal portion of
the **tunica vaginalis,** a serous membrane derived from
the peritoneum, which lines the scrotum and is reflected
upon the spermatic cord and testis.

The successive layers characteristic of the scrotal wall
are, therefore, from without inwards, the skin, the dartos
tunic, the spermatic fascia, the cremaster muscle, the in-

fundibuliform fascia, and the tunica vaginalis. Of these the dartos forms the basis of the median septum, while those internal to it invest the cavities which are separated thereby.

The **spermatic cord** of either side is composed of the vessels and nerves of the testis together with the duct of discharge, known as the **vas deferens**: they are imbedded in areolar tissue and surrounded by the continuations of the coverings of the testis. The structure of the duct will be described later.

The visceral portion of the tunica vaginalis invests the surface of the **testis** except its posterior border, along which it is reflected to become continuous with the parietal portion: it is frequently termed the **tunica adnata**. Beneath it lies the proper capsule of the testis, the **tunica albuginea**, a dense white fibrous layer of considerable thickness: along the posterior margin this is continued into the interior for some distance as a wedge-shaped fibrous reticulum known as the **mediastinum testis**, and also as the **corpus Highmori**: from the mediastinum radiate stout straight bands of fibrous tissue, the **septa** or **trabeculae**, which unite with the peripheral albuginea, thus imperfectly dividing the body of the testis into a number of irregularly pyramidal lobules. The inner stratum of the albuginea is quite vascular, and is sometimes distinguished as the **tunica vasculosa**: from it vascular trunks pass along the septa and form the blood supply of the lobules.

Each lobule consists of a variable number of seminiferous tubules supported by delicate lamellae of interstitial
tissue continuous with the stout fibrous septa as well as
with the proper membranes of the tubules: within these
lamellae peculiar epithelioid cells are found either singly or
in groups, whose origin and function are still uncertain;
they are known as the **interstitial cells** of the testis.

Each **seminiferous tubule** begins near the periphery of
the testis with an irregularly contorted portion extending
from its rounded extremity nearly to the mediastinum
and known as the **convoluted tubule,** or the seminiferous tubule in the strict sense: these occasionally branch
near their free extremities, and are said to anastomose in
some instances. As they approach the mediastinum they
become smaller and less irregular, and unite to form the
straight or **conducting tubules,** which lie in the apices of
the pyramidal lobules. On entering the reticular mediastinum the latter anastomose freely to form a network of
tubules of variable size, forming the **rete testis,** whose
meshes correspond with the spaces of the mediastinum.

The membrana propria of the convoluted tubules consists of several lamellae of endothelioid cells with flattened
oval nuclei. Upon this membrane rests an epithelial layer
several cells deep, from whose inner strata are derived the
male reproductive elements or **spermatozoa.** The basement membrane of the straight tubules is a continuation
of that of the convoluted: it is lined with a single layer of
cuboidal epithelium. In the rete testis the membrane becomes continuous with the reticular framework of the
mediastinum, and can no longer be distinguished: the

channels are lined with a single layer of flattened epithelial cells.

The arrangement of the epithelial cells of the convoluted tubules, and the changes which they undergo in the process of forming the spermatozoa, which is known as **spermatogenesis**, have been the subject of much controversy and cannot yet be said to be fully understood. Different methods have been described with positiveness not only in different classes of animals, but in different orders of the mammals, and even in different members of the same order. Our knowledge of the facts in the human subject are, as in most cases, less perfect than of those in the lower animals. In addition, the same terms have been applied by different writers to what are certainly different stages. The following account is based on that given by Schaefer in the tenth edition of Quain's Anatomy.

Next to the basement membrane is found a layer of cells most of which are cubical, clear, and possessed of nuclei which are in the resting or network phase: here and there dividing nuclei are seen: these cells may be termed **parietal cells** or **spermatogonia**. Scattered here and there among them are larger cells which project between the more internal layers: these are known as **sustentacular cells:** when much elongated and joined to bundles of developing spermatozoa they form the **columns of Sertoli.**

Within the parietal layer is seen a middle layer of somewhat larger spheroidal cells whose nuclei show various phases of division, the **intermediate** or **spermatogenic cells.** The layer may be one, two or more cells in depth;

its constituents are probably derived primarily from the cells of the parietal layer, but increase in number by lateral divisions. They are by some authors termed **spermatocysts.**

The inner layer of distinctly cellular elements consists of smaller and more numerous cells derived from the intermediate layer, the **spermatoblasts** (or **spermatids** of some writers): these are probably directly transformed into the spermatozoa. When first formed they compose a layer of closely packed small granular cells: they subsequently become more and more elongated vertically and collected into small groups, each of which becomes connected with one of the sustentacular cells above mentioned to form a column of Sertoli: later the bundle of spermatozoa derived from each group of spermatoblasts becomes separated from its sustentacular cell; the constituent elements are set free, and accumulate in large numbers in the lumen of the tubule.

The mature **spermatozoön** consists of an oval and flattened **head,** about four and a half micra long, two to three broad, and one to two thick; a cylindrical middle part or **body** about six micra long and less than one in diameter; and a tapering cilia-like **tail** forty to fifty micra long. The head consists chiefly if not entirely of the nucleus of the spermatoblast: the origin of the body and the tail is not so certain. In the lower animals the form and structure of spermatozoa vary greatly.

The pyramidal lobules and the body of Highmore, with their investment, the tunica albuginea, compose the whole of the gonad proper or testis in the strict sense. Closely associated therewith and often regarded as a portion thereof, though in reality the beginning of the efferent apparatus, is the **epididymis**, a tubular mass attached to the testis in the region of the mediastinum. Associated with the latter are certain rudimentary organs known respectively as the **hydatids of Morgagni**, lying between the head of the epididymis and the upper end of the testis, the **vas aberrans**, attached to the lower end of the epididymis, and the **organ of Giraldes**, found near the base of the spermatic cord.

The channels of the rete testis anastomose freely throughout the whole of the mediastinum or body of Highmore. Those of the upper or anterior portion open into the **efferent tubules**, also known as **vasa efferentia**, from twelve to twenty in number, which penetrate the albuginea and enter the epididymis, of which they form a part: at first straight, they soon become coiled in conical masses, the **coni vasculosi**, whose bases are turned away from the testis. These aggregated conical bodies, together with the upper portions of the **canal of the epididymis**, into which the efferent tubules open, form the **globus major**, or **caput epididymis.** Their constituent tubules are as large as the convoluted tubules of the testis: they are lined with a cuboidal or short columnar ciliated epithelium, beneath which is a well defined membrana propria surrounded by a thin transverse layer of smooth muscular fibres.

The **canal of the epididymis**, exceedingly flexuous from the point of its origin in the extremities of the uppermost vasa efferentia, becomes disposed in numerous small irregular coils in the middle region and is continued at the lower portion as a densely convoluted mass, the **globus minor**. At its origin it is twice the diameter of the vasa efferentia: lower it becomes smaller, and is enlarged again in the globus minor, from which it is continued as the beginning of the **spermatic duct** or **vas deferens**. It is lined throughout its course with tall columnar cells (between whose bases smaller rounded cells occur): these cells are provided with unusually long cilia throughout the greater portion of the tube, the cilia disappearing in the lower portion. Beneath the underlying membrane is a thin circular layer of smooth muscular fibres continuous with that of the vasa efferentia, which is surrounded by a thin longitudinal layer. The convolutions of the tube are bound together by interstitial areolar tissue which is here and there replaced by incomplete fibrous septa which partially divide the 'epididymis into irregular **lobules.**

The **hydatids of Morgagni** are small saccular bodies lying between the globus major and the upper end of the testis. One of these, the **stalked hydatid,** is provided with a short peduncle, is usually present, and has a homologue in the female organs of reproduction. The others, the **sessile hydatids,** are variable in number and sometimes wanting: they are found in the male only. Both stalked and sessile hydatids are lined with cuboidal epithelium upon which cilia have been described.

The **vas aberrans** is a long narrow blind tube, a diverti-

culum of the canal of the epididymis, which arises from
the latter at or near the point where it becomes continuous with the vas deferens. Like the canal, it is exceedingly
tortuous in its course, forming an elongated convoluted
mass which extends upward among the vessels of the
spermatic cord. It resembles the vas deferens in structure. It is almost invariably present, is sometimes branched, and in some cases more than one such structure occurs.

The **organ of Giraldes,** also called the **paradidymis,** is a
small body found on the front of the spermatic cord just
above the globus major. It consists of several disconnected irregularly branched tubules lined with columnar
ciliated epithelium. Their coiled masses form irregular
nodules imbedded in the connective tissue of the regions
between the cord and the epididymis.

The **spermatic duct,** or **vas deferens,** is the continuation of the canal of the epididymis. It is considerably
larger and more complex in structure than the canal,
showing like many of the tubular structures of the body,
a definite mucosa, submucosa, and musculosa. The mucosa consists of a stout membrane sometimes thrown
into longitudinal folds, and bearing a non-ciliated columnar epithelium: the submucosa is composed of areolar tissue somewhat laminated in arrangement: the musculosa
is thick and yellowish in color and comprises an inner circular and an outer longitudinal layer of smooth muscular
fibres, external to which is a fibrous adventitia. At the

commencement of the duct there is also a longitudinal layer of muscular fibres internal to the circular layer.

The **ampulla**, or sacculated enlargement of the duct situated near its junction with the seminal vesicle, resembles the rest of the duct in structure, save that the various coats are somewhat thinner. The tubular diverticula of the vasa deferentia known as the **seminal vesicles** are like the ampullae in structure. The spermatic duct or vas deferens of either side unites with its associated seminal vesicle to form the **ejaculatory duct**, or **common seminal duct**, which completes the passage way from the seminiferous tubules of the testis to the urethra: in this region the walls of the tube are much thinner than in the vas deferens and the external fibrous adventitia disappears as the duct enters the substance of the prostate gland.

The **male urethra**, into which the common seminal ducts empty within the region surrounded by the prostate gland, is by virtue of that fact divisible into two distinct regions, the **urinary** and the **urino-genital**: it is the former alone that corresponds to the female urethra. By its anatomical relations it is also divided into the **prostatic**, the **membranous**, and the **penial urethra**: the former includes the first of the two regions above mentioned and a portion of the other. Histologically the structure differs in the several regions chiefly in the character of the epithelium, in the special structures found in the mucosa of the penial region, and in the composition of the musculosa.

The **prostatic urethra** of the male resembles the female urethra in being essentially a continuation of the wall of the bladder. It has a mucosa lined with transitional epithelium, which rests upon a membrane rich in elastic fibres. Beneath is a highly vascular submucosa, and beyond this a musculosa which consists of an inner longitudinal and outer circular layer of smooth muscular fibres. In front of the openings of the common seminal ducts the epithelium passes by a 'gradual modification from the transitional to a somewhat stratified columnar form. The **membranous urethra** continues the structure of the anterior portion of the prostatic, and in addition receives a distinct layer of striped muscular fibres from the adjacent compressor urethrae.

In the **penial urethra** the epithelium is composed of a single layer of columnar cells except at the fossa navicularis, where it passes into stratified squamous epithelium continuous with that of the surface of the glans. Here and there the mucosa exhibits irregular depressions of variable size, known as **lacunae Morgagni**: it also has connected with it numerous small racemose glands, the **glands of Littre,** which are also sparingly found in other portions of the urethra; they are lined with cuboidal or low columnar glandular cells.

The **penis** consists essentially of three masses of what is commonly called erectile tissue, the right and left corpora cavernosa, and the median inferior corpus spongiosum; the latter being traversed throughout its length by the penial urethra, and having its distal portion expanded to form the glans penis: the whole, of course, invested by the somewhat modified integument. The form and rela-

tions of these bodies are matters for anatomical discussion: we are here concerned with their histological structure only.

What is known as **erectile tissue** consists primarily
"simply of a somewhat circumscribed collection of larger
and smaller veins which under certain circumstances may
become distended with blood, thus causing the parts in
which they lie to expand." Its specialization consists
chiefly in the enlargement of the veins to form irregular
sinuses, or **cavernae,** and their frequent anastomosis; the
development and modification of the circumscribing skeletal tissues; and accessory modifications of the arteries
of supply. That the cavernous sinuses are to be regarded
as veins is shown not only by the structure of their walls,
but also by the fact that the blood reaches them chiefly if
not entirely through capillaries. It will be seen from what
has just been said that erectile tissue is not constant in
structure as are adenoid and other compound tissues that
have been previously described. The form assumed in
each locality where it occurs will therefore be briefly
stated.

The **corpora cavernosa** are surrounded and united by a
stout fibrous envelope, termed, like that of the testis, the
tunica albuginea; it consists of bundles of white fibres
chiefly disposed in a longitudinal direction and mixed with
numerous elastic fibres: within, the white fibres are chiefly
circularly disposed, surrounding each of the two corpora
to form an **individual sheath**; these two sheaths are

confluent in the mid-plane through the greater part
of the penis, forming a septum which is incomplete by
virtue of the presence of slit-like apertures through which
the erectile tissue becomes continuous from side to side:
these apertures are most numerous in the anterior part of
the penis.

From the fibrous sheath numerous stout **trabeculae**
pass inward to form a reticulum whose meshes are the
cavernous sinuses. They are composed chiefly of white
fibrous tissue, with more or fewer elastic fibres intermingled, and contain in addition numerous bundles of
smooth muscular fibres: their surfaces are lined with the
vascular endothelium of the sinuses.   The trabeculae are
stoutest and most numerous near the surface: the sinuses
are correspondingly largest at the centre of the body;
they are also larger near the extremity of the penis than
at the base, in which region their long diameter is as a
rule placed transversely to the penis.   The small arteries
which follow the trabeculae in many cases project from
their surfaces in peculiarly curled and coiled loops; they
are hence known as **helicine arteries**: they are said in
some cases to open directly into the sinuses, but such a
direct communication between an artery and a modified
vein is not easily demonstrated under conditions which
exclude the possibility of error.

The **corpus spongiosum** differs from the corpora cavernosa histologically in the thinness and increased elasticity
of its fibrous tunic, which contains so much elastic tissue
as to be yellowish in color; in the smaller size and greater

uniformity of the trabeculae; and in the lesser amount of muscular tissue which they contain. The venous sinuses are smaller and more uniform, forming a spongy mass, from which the name is derived: their greatest dimensions are as a rule longitudinally disposed.

In the **glans** the meshes are quite small and uniform in size: the erectile tissue passes insensibly into the lower strata of the integument, with which its surface is invested. The derma is thin and highly vascular over the surface of the glans, and the epidermis has the form of a stratified squamous epithelium devoid of the division into layers characteristic of the cuticle and resembling in character that found on the oesophageal mucous membrane. Glands are wanting, except upon the corona and the cervix, where modified sebaceous glands, the **glands of Tyson**, are abundant. Special nerve terminals, the so-called **genital corpuscles**, are present, as are Pacinian bodies.

The **skin** of the penis is quite thin, highly elastic, and very movable, and contains but a very little fat: the larger portion is devoid of hair also: as it passes around the free margin of the prepuce it changes its structure and the character of its epithelium, the lining of the prepuce, like the investment of the glans, having the appearance of a mucous membrane. At the base it passes on into that of the pubes, which is quite thick, beset with coarse hairs, and provided with a dense fatty layer.

The urino-genital tract of the male has associated with it certain glandular bodies, the **prostate gland**, which surrounds the proximal region of the urethra, and the **glands of Cowper,** paired organs opening into it near the point where it enters the corpus spongiosum. Opening into it ventrally in close proximity to the apertures of the ejaculatory ducts is an interesting rudiment, the **sinus pocularis,** otherwise known as the **uterus masculinus.** These structures will next be described.

The **prostate** is a glandular body which differs from most organs of the kind in the fact that not only its capsule but also its stroma contains a very considerable amount of smooth muscular tissue. This muscular tissue is in continuity with the musculosa of the urethra and of the ejaculatory ducts, and posteriorly with that of the bladder. The capsule is divisible into two layers, between which is found the prostatic venous plexus: from it trabeculae pass inward to form the framework of the gland, consisting, in addition to the smooth muscular tissue already mentioned, of a very small amount of white fibrous tissue and a larger quantity of elastic fibres. The alveoli are tubular, frequently quite elongated and irregular in shape, their walls sometimes showing conspicuous folds: the epithelium is columnar and simple save that frequently small and spheroidal cells are found at the base of the columnar cells. The ducts, which are numerous, are lined with columnar epithelium which changes into stratified as it approaches their openings upon the urethra.

Cowper's glands are small bodies of the racemose type, each consisting of several small lobes. Their capsules and supporting framework resemble those of the prostate to some extent in the presence in each of a small amount of smooth muscular fibre: a well defined longitudinal layer of smooth muscular fibres is also present in the wall of the principal duct. The acini resemble those of a mucous salivary gland in form and in the general appearance of the glandular epithelium: there is a conspicuous lumen, the cells are pyramidal, and the nuclei are situated near the base. Nothing resembling the crescents or demilunes of the mucous glands has been observed. The lobar ducts are lined with cuboidal epithelium, which passes into columnar in the principal ducts.

The **sinus pocularis** is the homologue in the male of the vagina and uterus of the female. It is a diverticulum of the prostatic urethra having a well-defined muscular wall and a mucosa containing a number of short tubular glands which resemble the uterine glands in their form and structure.

CHAPTER XIX.

THE FEMALE REPRODUCTIVE ORGANS.

———

The female reproductive apparatus consists of the female gonads, or **ovaries,** in which the reproductive elements are formed, the oviducts, or **Fallopian tubes,** by which they are conveyed from the ovaries; the **uterus,** in which they are received, and in which the fertilized ovum or oosperm develops into the embryo; the **vagina,** by which the uterus communicates with the exterior, and the parts composing the **vulva,** which immediately surrounds the opening of the vagina. As in the case of the male, there are also present certain rudimentary bodies, chiefly in the vicinity of the gonad.

The **ovary,** like the testis, is an organ in which specialized cells, epithelial in their origin, are matured and liberated: the sexual elements differ, however, in an antipodal manner as regards their size, activity, destination, mode of transportation thereto, and mode of liberation: there is a corresponding difference in the structure of the organs which produce them, the ovary having nothing of that tubular structure seen in the testis and giving to that body a close resemblance to a gland.

The framework or **stroma** of the ovary lacks the abundant white fibrous tissue found in the capsule and trabeculae of the testis: it consists chiefly of a peculiar form of connective tissue characterized by the presence of elongated nucleated cells which are frequently spindle shaped, and by the scarcity of true fibrous tissue, either white or elastic. It contains numerous smooth muscular fibres, which are most abundant in the deeper portions. Toward the surface the stroma becomes more dense, forming quite a well-defined superficial layer, to which the name of the **tunica albuginea** has been given; it lacks the firmness and definiteness of the layer so designated in the testis. At the base of the ovary the stroma is especially rich in blood vessels: the region occupied by them is known as the **zona vasculosa.**

The region between the albuginea and the zona vasculosa constitutes the **parenchyma** of the ovary. It is rather indefinitely divided into a cortical and a medullary portion by the character of the **Graafian follicles** contained in it. It also contains, scattered irregularly through it, groups of **interstitial cells** similar to those found in the testis.

The surface of the ovary is invested with a layer of cells which are structurally continuous with the serous endothelium of the peritoneum, but which differ therefrom in form and function. They are cuboidal or low columnar in shape, and constitute the **germinal epithelium:** they have no proper basement membrane, but rest directly upon the tunica albuginea. Here and there may be seen, especially in the embryo, certain cells which are larger and more rounded in form: these are the **primitive ova.** These, during foetal life and possibly in childhood sink into the stroma, accompanied by tubular or spheroidal nests of epithelial cells: it is doubtful whether this ever takes place in the adult: it is also at present a matter of question whether the primitive ova increase in number by division after they have passed into the stroma, or whether all so situated have come from the epithelium of the surface of the ovary.

The cortical region of the parenchyma is crowded with the spheroidal masses of cells formed in the manner just described, each consisting of one or sometimes two primitive ova surrounded by a layer of epithelial cells. These are the primitive **Graafian follicles.** In those immediately beneath the albuginea the surrounding layer is usually but a single cell deep: at first flattened, and hardly distinguishable from the cells of the adjacent stroma, its cells soon become cuboidal in form. At the same time the fibres of the stroma tend to assume a disposition concentric to the follicle, forming the beginning of the **theca folliculi.**

As the follicles grow older they tend to sink deeper into the stroma of the ovary; the cells of the enveloping layer at the same time proliferating, and the layer becoming several cells thick. Shortly afterward the ovum leaves its central position for one nearer one side of the follicle, usually that farthest from the surface of the ovary, while a cleavage takes place in the cellular layer toward the other side, the space formed becoming infiltrated with a clear fluid, the **liquor folliculi.** The follicles now rapidly increase in size, at the same time sinking into the medullary region, where they are seen as large vesicles filled with fluid, invested by a well-defined theca, and lined by a layer of small isodiametric cells of irregular form, the layer being several cells deep: it is now known as the **membrana granulosa.** Attached to it at one side of the follicle is the rounded heap of similar cells which contains the ovum: it is termed the **discus** or **cumulus proligerus.**

The follicles still increasing in size, their outer wall now tends to approach the surface of the ovary, the fully matured follicle finally projecting somewhat from the surface, by whose rupture the contained ovum is eventually to be liberated. The theca is now well defined and consists of two layers, an inner or vascular, and an outer or fibrous layer. The cells of the membrana granulosa next the theca and those of the cumulus next the ovum are distinctly columnar in form. The most conspicuous as well as the most important body present is the **mature ovarian ovum.** This is a spheroidal body now much larger than the primitive ovum from which it was developed, although small as compared with the ova of many of the lower vertebrates. It is in the human subject about two-tenths of a millimetre in diameter. It is invested by a thick covering appearing when seen with microscopes like those used by the earlier observers to be quite clear: as its optical section forms a girdle or zone of considerable breadth about the ovum it was named by Von Baer the **zona pellucida.** Careful examination by modern instruments and methods demonstrate that it contains innumerable radial striae: it is therefore now commonly called the **zona striata,** or still more accurately the **striated membrane.**

Several eminent and accurate observers have described a
delicate membrane, which they call the **vitelline membrane,** internal to the layer just described: its presence
cannot be readily demonstrated with certainty.

Within the envelope just described is the **vitellus,** or
yolk, wrongly so called, since it does not correspond to
the body so termed in the eggs of many lower vertebrates:
it is a homogenous protoplasmic mass, semi-fluid in consistency and highly granular. It contains, usually in an
eccentric position, a large spherical nucleus which was
named by Purkinje the **germinal vesicle:** the nuclear contents exhibit a coarse network characteristic of the phase
of complete rest: there is usually but a single nucleolus,
which is quite large and rounded, and was called by Wag-
ner the **germinal spot.**

When the ovum is discharged by the rupture of the follicle upon the surface of the ovary, the follicular cavity is
at first filled with a clot of blood. It is quickly invaded
by growths from the wall of the follicle formed in part of
rapidly proliferating cells of the membrana granulosa, in
part of folds and processes from the theca: it has been as-
serted that the interstitial cells previously mentioned also
enter to a large extent into the ingrowing structure. The
result is the formation (about the shrunken and discolored
clot as a centre) of a mass of mingled cells and fibres known
as a **corpus luteum:** this is at first sharply defined by the
presence of the theca, but gradually loses its definiteness,
and becomes continuous with the mass of the ovary, the
peculiar spurious tissue thus formed composing quite a

large part of that organ in age. The corpus luteum formed
concurrently with pregnancy is large and well-defined and
is regarded as characteristic: but corpora lutea equally
large and distinct sometimes (though more rarely) occur
under other conditions.

The **oviducts**, commonly termed the **Fallopian tubes**,
while they vary in form in the several regions distinguished by the anatomist, are quite uniform in their histological structure throughout their whole extent. Continuous
at the isthmus with the uterus, they open at the fimbriated
extremities into the peritoneal cavity: they consequently
present the only instance of direct continuity between a
mucous and a serous surface. Like nearly all the larger
tubular structures in the body, the wall is divisible into a
mucosa, a submucosa and a musculosa; to which is added
a serosa derived from their investment by the marginal
fold of the broad ligament.

The **mucosa** consists of a well developed fibrous membrane moderately rich in elastic fibres, and well supplied
with blood vessels and lymphatics, which supports a layer
of simple columnar ciliated epithelium. An imperfectly
developed muscularis mucosae, consisting of longitudinal
bundles of smooth fibres, is also present. The mucosa
throughout its extent is thrown into longitudinal folds
which in the ampulla and particularly in the infundibulum
are very extensive and have secondary folds upon their
surfaces, giving to the cross section a peculiar arborescent
appearance. As seen in such sections the bases of these
folds often present the appearance of tubular glands; but

true glands are not present. The inner surface of the fimbriae is lined with the mucosa, while the outer surface is covered with the serosa: the two becoming confluent along the sides.

The **submucosa** is a simple layer of areolar tissue of but slight depth. It is continuous with that of the uterus, like which it contains small ganglia and scattered multipolar cells, forming the rudiments of a plexus. The **musculosa** consists of an inner circular and an outer longitudinal layer of smooth muscular fibres, the latter being but imperfectly developed. The **serosa** consists of a thin fibrous membrane supporting the serous endothelium characteristic of the surface of the peritoneum.

Attached to the extremity of the tube or to one of the fimbriae is frequently found a **pedunculated cyst** or **stalked hydatid of Morgagni.** It is the homologue of the body bearing the same name in the male reproductive apparatus and resembles it in general structure, the cavity of the sac and also of the pervious portion of the stalk being lined with cuboidal or columnar epithelium.

Situated in the broad ligament between the ovary and the ampulla of the oviduct is a well-defined mass of irregularly convoluted tubules, known as the **parovarium.** It is also called, from its discoverer, the **organ of Rosenmueller:** Waldeyer has proposed for it the name of the **epoophoron.** The constituent tubules are lined with low columnar epithelium: they converge toward each other, without uniting, at the ends nearest the ovary: the other

extremities diverge somewhat, and terminate in a longitudinal tube which runs parallel with the oviduct : in some
of the lower mammals this tube is quite extensive and is
known as the **duct of Gartner,** a term also applied to it
in the human subject by many. Somewhat nearer to the
uterus than the parovarium a smaller and more irregular
group of rudimentary tubules occurs, similar in structure
to those just described. These have been designated by
Waldeyer the **paroophoron.** The homologies of these rudiments will be discussed later.

The **uterus** presents but two distinct regions histologically, the **fundus** and **body** agreeing in structure and differing from the **cervix.** The most characteristic features of
the upper region are found in the **mucosa,** the stroma of
which is greatly modified, while the muscularis attains a
greater development than does the structure bearing that
name in any other portion of the body. The surface is invested with a single layer of columnar ciliated cells directly continuous with that lining the Fallopian tube. Beneath this
is a very thick mucous membrane containing but a very
small quantity of fibres and composed in large part of
spindle shaped cells similar to those found in the stroma of
the ovary, loosely interwoven : the spongy mass so formed
contains numerous lymph spaces and leucocytes. Imbedded in it are great numbers of tubular **uterine glands,**
wavy or convoluted in their course, not infrequently
branched, and penetrating to the base of the stroma and
quite frequently between the bundles of fibres of the muscular layer : they are bounded by a delicate basement mem-

brane which supports columnar cells similar to those lining the uterine wall: near the blind extremity of the tube
the columnar cells entirely fill its cavity: but throughout
the greater part of its extent there is a distinct lumen.

The muscularis mucosae is the chief muscular coat of the
uterine wall; greatly developed at all times, it is enor-
mously hypertrophied during pregnancy, partly by the
great increase in number. It consists of bundles of fibres
interwoven with a sparing amount of interstitial connective tissue, and running in various directions: their disposition is apparently quite irregular, and cannot be described briefly with clearness: it is, moreover, subject to
considerable variations: it can perhaps be best understood
by regarding it as consisting chiefly of circularly disposed
bundles which are arranged on the fundus in two sets, one
concentric to the insertion of each of the two oviducts,
and which become gradually combined to form a single
set as they approach the lower extremity of the body.

There is associated with the great development of the
muscularis mucosae a corresponding reduction of the submucosa, there being less independent movement of the
mucosa and musculosa in this case than in almost any of
the similar hollow structures. There is, however, a distinct zone of connective tissue discernible just exterior to
the muscularis mucosae characterized particularly by the
presence of numerous bloodvessels and lymphatics, and
by scattered nervous elements. It is of such slight extent
as to be by some described as wanting.

The **musculosa** consists of two distinct layers of smooth
fibres both quite thin and varying in their relative devel-

opment in different parts of the organ. The inner or circular layer is the more uniform in thickness and in the arrangement of its fibres: the fibres of the outer layer are in the main longitudinally disposed, but are somewhat irregularly arranged upon the fundus, in accordance with its irregularities of form. This layer also gives off bundles of muscular fibres extending out into the ligaments of the uterus. The musculosa is invested by a **serosa** which is a continuation of the peritoneum.

⸰ The **cervix** differs from the region just described chiefly in the structure of the **mucosa,** the stroma of which is much firmer and richer in fibres, both white and elastic, the membrane being thrown into characteristic folds, and in the lower portion beset with minute papillae. The upper two-thirds is lined with columnar ciliated epithelium continuous with that of the upper regions: this passes below into the stratified squamous epithelium which invests the papillated lower third. There are present both tubular and saccular glands said to be lined in each case with columnar ciliated epithelium: the saccular glands contain also goblet cells and secrete the thick mucus found in the cervix. Here and there spheroidal bodies filled with a clear yellowish fluid can be seen with the naked eye: they are probably occluded and enlarged mucous glands, and are known as **ovula Nabothi.** The muscularis mucosae is well developed and consists chiefly of circular bundles: these are accumulated in greater numbers at t⸱ upper and lower extremities of the cervix to form the sphincters of the regions in question.

The **submucosa** is not conspicuous. The **musculosa** resembles in structure the same portion of the wall of the body of the uterus: the inner circular and outer longitudinal layers are clearly defined. The region of the cervix toward the rectum is invested with a peritoneal **serosa**: that toward the bladder is separated from that organ by an adventitia of areolar tissue. The portion of the cervix that projects into the vagina to form the **os uteri** agrees in structure on its inner aspect with the cervix: on its outer it is a continuation of the vaginal wall.

The **vagina** differs from the uterus in structure in accordance with its differing and double function, it serving at once as the channel by which the male reproductive elements are brought into proximity with the female elements, and as the avenue of discharge for the foetus at its maturity. Its wall is muscular, dilatable, and highly elastic, somewhat erectile, and provided with a definite reïnforcement of adenoid tissue; a feature possessed by no other portion of the reproductive tract.

The **mucosa** of the vagina is lined with stratified squamous epithelium, which rests upon a thick mucous membrane. The surface of the latter is beset with minute papillae which project into the deeper portions of the epithelium but do not produce a noticeable roughness of its surface, the only irregularities observable being those due to the well-marked folds or **rugae**. The upper portion is a dense fibrous layer rich in elastic fibres: below this in the rugae are networks of large veins supported by fibrous tissue containing numerous bundles of smooth muscular fibres which may be regarded as representing the muscularis mucosae, elsewhere absent: a ridge of rudimentary erectile tissue is thus formed beneath each ruga. Leucocytes abound in the mucosa, and scattered nodules of adenoid tissue are found: in the anterior wall near the orifice there is a well-defined adenoid layer. Special nerve terminals, the **genital corpuscles** of Krause, as found in the mucosa of the vagina. It is doubtful whether glands of any sort are present.

The **submucosa** is quite loose in structure and contains a venous network whose meshes run chiefly in the direction of the vagina. Beyond the submucosa is the **musculosa**, which is not sharply defined, as in most cases, into distinct strata: the inner bundles are in the main circularly disposed, and the outer bundles longitudinally; the two regions being, however, blended by numerous oblique bundles. A well-marked **fibrosa** invests the musculosa: it is composed largely of elastic tissue and is best developed on the anterior wall: it also contains an extensive plexus of large veins intermingled with bundles of smooth muscular fibres and forming a layer of erectile tissue best developed near the lower extremity.

The **hymen** agrees in its structure with a fold of the mucosa of the vagina, and can perhaps be regarded as derived therefrom, though its presence in rare instances in cases of absence of the vagina has caused this mode of origin to be questioned by those who regard it as a fold of the skin of the vestibule.

The **vulva** includes a number of parts or regions each characterized by certain histological features worthy of brief mention. The surface of the area known as the **vestibule** is covered by a mucous membrane continuous with that of the vagina and of the urethra at their respective orifices. It is covered with stratified squamous epithelium, contains numerous elastic fibres, and is feebly erectile: it also contains numerous simple mucous glands. At the lower limit of the vestibular area are seen on either side of the vaginal orifice the openings of the ducts of the **glands of Bartholin**, small racemose glands of the mucous type homologous with the glands of Cowper in the male subject. Beneath the mucous membrane of the vestibule and somewhat external to the proper limits of the vestibular surface are paired elongated masses of erectile tissue, the **bulbi vestibuli**, whose converging upper extremities are continuous with smaller plexuses whose vessels are confluent above with those of the glans clitoridis. The bulbar regions may be regarded as corresponding to the bilateral bulbous portions of the corpus spongiosum of the male subject.

The **clitoris** consists of two small corpora cavernosa identical in structure and relations with those of the male, and a small glans of spongy erectile tissue, which is of course imperforate. Its surface contains numerous genital corpuscles. Its preputial fold is continuous with the upper and its fraenum with the lower of the anterior divisions of the **labia minora.** The latter, while resembling the surface of the vestibule in color and texture, may be regarded as folds of the skin: they contain numerous large sebaceous glands, but sweat glands are wanting, as are also hairs: their inner surface contains numerous genital corpuscles. Fat is wanting in the subcutaneous connective tissue, but large irregular venous channels are present, with smooth muscular fibres, composing here as elsewhere a loose form of erectile tissue. The **labia majora** are well defined folds of the skin whose inner surfaces resemble in appearance the outer surfaces of the labia minora, with which they are confluent, but differ from them in the presence of occasional modified sweat glands and minute hairs, associated with the sebaceous glands common to both. The thick and rounded margin of the fold affords a gradual transition from the moist stratified'squamous epithelium of the mucous type found upon the greater portion of the vulvar surface to the epidermis of the skin, with which its outer surface agrees in general structure. In the deeper portions of the integument of the labia majora is found a layer of tissue similar to that forming the dartos tunic of the male scrotum, with which the labia correspond. Like the divisions of that structure they converge above to be united upon the pubic eminence in the region known as the **mons veneris**, characterized, as in the male, by an abundance of coarse curling hairs, of numerous enlarged sudoriparous glands, and by a dense mass of subcutaneous fat.

The **mammary glands**, while essentially tegumentary organs, differ so greatly from all other dermal glands as to require consideration apart from the latter: their functional relations to reproduction render it appropriate to discuss them at this time. Usually but not always rudimentary in the male subject, they are normally but not invariably fully developed in the female. Each mammary gland so called is in reality an aggregate of fifteen or twenty distinct glands, if we regard that term as strictly designating a secretory body provided with a duct which opens independently, since each of the ducts opens by a separate orifice upon the skin of the nipple: but their union into a single anatomical structure is so intimate that it is more convenient to designate each of these regions as a lobe of a compound racemose gland. The structure of these lobes and of the whole organ varies greatly in relation to functional activity; but it is characteristic of it at all times that it possesses an unusually large proportion of connective tissue and fat in its composition.

Before the gland has been called into functional activity the lactiferous ducts are present, as well as the ducts of the lobules by whose confluence they are formed: the latter have at the extremities rudimentary acini, which are, however, relatively few in number, and consist of masses of epithelial cells. Acini, lobules, and lobes are alike imbedded in an extensive stroma of connective tissue which forms stout septa not only between the lobes but between their subdivisions as well; while a considerable amount of subcutaneous and interstitial adipose tissue is present.

As pregnancy advances the acini become larger and more numerous, still consisting, however, of solid masses of cells. At the time of delivery the central cells undergo fatty degeneration and form the **colostrum corpuscles** of the milk of commencing lactation. In the fully active

gland the acini are spheroidal, comparatively large, and
are lined by a single layer of cells which when at rest are
flattened, but during secretion become cuboidal or columnar in form, their extremities containing one or more large
oil-droplets in each instance: the latter are liberated by the
rupture of the cell substance, a portion of which is probably contributed to the secretion. The basement membrane upon which the epithelium rests consists of an endothelioid layer of connective tissue corpuscles: the intralobular stroma is greatly reduced in proportional quantity. The terminal branches of the ducts have a thin basement membrane like that of the alveoli, which is lined by
a single layer of flattened cells, their appearance resembling that of the ductules of the salivary glands: the larger
lactiferous ducts have stouter walls, and cuboidal epithelium; these empty into the still larger channels, the **galac-
tophorous ducts,** one of which leads, as has been stated,
from each lobe to an independent orifice upon the skin of
the nipple. These large ducts have stout walls of fibrous
and elastic tissue, containing a few smooth muscular
fibres: they are lined with columnar epithelium save in
their outermost portions, where the epithelium becomes
stratified. Each has an enlargement, the **ampulla,** near
its termination at the nipple, whose structure does not
differ from that of the rest of the duct.

In the intervals between lactation the mammary glands
assume a resting condition in many respects similar to
their primitive state. They always contain, however, a
smaller amount of dense fibrous tissue and a larger
amount of fat, and are consequently much less firm in

texture. At the close of the reproductive period they begin to undergo a retrograde metamorphosis, the acini and smaller ducts disappearing, the larger ducts collapsing, and the shrunken organ consisting chiefly of a mass of connective tissue and fat.

The **nipple** is a cylindrical projection from the surface of the gland, covered with deeply pigmented skin and composed of the extremities of the lactiferous ducts, associated bloodvessels, and smooth muscular fibres arranged in circular and longitudinal bundles. The dermal papillæ are rich in nerve terminals: there is no subcutaneous fat: scattered in the surface are the small racemose **glands of Montgomery**: the areola at its base contains sweat-glands and numerous sebaceous glands.

The intimate relations existing between the urinary and reproductive systems and the homologies between the male and female sexual organs cannot be clearly stated without a description of their formation in the embryo, accompanied by a statement of some at least of the facts of their comparative anatomy. No attempt will be made at this time to discuss systematically either the embryology or the morphology of these organs, but such account of each will be alone given as appears necessary to the intelligent comprehension of the relations and homologies above referred to.

The structure fundamental to the whole of the urogenital apparatus is what is known as a **nephridium.**

In the great majority of the classes of the higher invertebrates the elimination of nitrogenous waste products
takes place through the agency of organs designated
by that term. A nephridium is essentially a tubular
structure opening at one extremity upon the surface of
the body (either directly or indirectly), and at the other
communicating with the body cavity, or **coelom**, by a
more or less funnel-shaped extremity termed the **ne-
phrostome.** Its wall is lined with an epithelium which
is glandular throughout a large portion of its extent:
the funnel-shaped internal opening is commonly and the
larger portion of the rest of the tube frequently ciliated, the ciliary movement invariably sweeping toward
the external opening. The tube may be simple and
quite direct in its course, or long, convoluted, and divided into specialized regions. Where a well developed
vascular system is present the more or less coiled nephridial tube is usually provided with a rich net-work
of capillaries. Nephridia may be simple: a single pair
opening right and left on the surface of the body, as in
many mollusks; or a pair being found in each of the
majority of the segments of the body, as in the annelids,
where they were first observed, and designated from
their disposition **segmental organs**: or they may be
compound, a number of funnel-bearing and ciliated tubules opening symmetrically into lateral tubes which
discharge {the secretion of the tubules upon the surface
of the body or into a posterior cloacal sac, as in the roti-
fers. Without entering into a discussion of the homologies that may exist between the nephridia of the various

classes of invertebrates, or between those of either of
them and those of the vertebrates, it may be said that the
urinary apparatus of the latter conforms, in a general
way at least, to the plan last described.

The gonads, in which the reproductive elements are
formed, are in most classes of Metazoa developed within
the body-cavity, or **coelom**, into which they protrude,
and into which the reproductive elements, when fully
mature, are in many species discharged. In some of the
lower forms where this takes place, and in which the reproductive elements are produced in immense numbers,
the body wall becomes distended by their presence and
finally liberates them by rupturing, the parent organism
being thereby destroyed; a condition comparable to that
of the so-called annual plants, which grow, blossom, ripen a
single crop of seeds, and die. In other cases the nephridia serve as channels for the escape of the reproductive elements: it will be noted that the small and actively
moving spermatozoa would pass far more readily
through the nephridial tubules than the large and passive
ova; and it is probably true in the case of some species
that the female is destroyed by the rupture of the body
wall after the ripening of the first crop of ova, while the
male elements are liberated through the nephridia, the parent organism surviving. In most of the higher classes of
animals there are distinct channels of discharge, the **gon-
aducts**, for the reproductive elements: these are in some
cases clearly formed by the modification of nephridia.

The vertebrate urinary apparatus consists in effect of a
series of nephridial tubules arranged along the dorsal wall

of the body cavity on either side of the mid-plane and
opening into right and left ducts of discharge. The disposition, form and relations of the tubules, and the origin
and position of the duct of discharge of the functional kidney undergo important modifications in the various
classes of vertebrates. Taking the group as a whole the
series of tubules may be said to be divided into three regions, the **pronephros** or **head-kidney**, the **mesonephros**
or **primitive kidney**, and the **metanephros** or **permanent
kidney** of the higher vertebrates. Each of these regions
needs brief consideration.

The **pronephros** is best developed in the anamnia (cyclos-
tomes, fish-like vertebrates, and amphibia), in all of which
it is present as a well-defined structure in the embryo, and
in some of which it is functional in the adult. The tubules
are always few in number, sometimes but one on each
side, open upon the body cavity by distinct nephrostomes,
and discharge into ducts leading to the cloaca and known
as the **segmental ducts**. Near the opening of the nephrostomes there is found on each side a process of the coelo-
mic wall containing a mass of capillaries and constituting
a prominent glomus: the region of the coelom where this
occurs is either partially or in some cases wholly constricted off from the rest of the body-cavity so as to form
"what is practically an enormous Malpighian body:"
thus foreshadowing the arrangement found in the permanent kidney of the mammals. The development of the
pronephros is embryonic and rudimentary in the sharks
and their allies among the fish-like vertebrates, and in all
amniota (reptiles, birds, and mammals): in the last named

class it is doubtfully represented in the adult male and female by rudimentary structures connected with the
surprarenal capsule.

The **mesonephros** is the functional kidney of all anamnia except the sharks and their allies. In this class and in
the various classes of the amniota it is represented by the
embryonic structure termed the **Wolffian body**. It consists primarily of serially arranged tubules plainly homologous with those of the pronephros, although their mode
of development is not identical, being somewhat acceler-
ated; a point of importance in connection with the origin
of the metanephros.   These tubules become variously
coiled and convoluted, and may give rise to branches like
themselves: in many anamnia they open upon the peritoneal surface by nephrostomes: they have associated
with them glomeruli resembling in their origin the glomus just described in connection with the pronephros: in
some cases several tubules open upon a single glomerulus.
They discharge into a modification of the segmental duct
known as the **mesonephric** or **Wolffian duct**: it differs
from the primary segmental duct in undergoing (in some
species) longitudinal cleavage to give rise to a second
canal called the **Muellerian duct.**

The ripened gonads of many anamnia discharge the reproductive elements into the body cavity by rupture,
from which they escape through openings posteriorly and
ventrally placed and known as abdominal pores. In
other cases abdominal pores are wanting: in the female
the ova pass out through the Müllerian duct or **oviduct**;
in the male the testis lies opposite the anterior end of the

primitive kidney, or Wolffian body; diverticula grow out from the Wolffian tubules of that region and become connected with the testis, forming channels of escape for the spermatozoa, which thus reach the Wolffian duct; the latter therefore functions both as a ureter and a **spermiduct**.

In the selachians (the sharks and their allies), and in all amniota a diverticulum is given off from the posterior portion of the Wolffian duct. In the former class this becomes connected with a mass of serially disposed tubules arising posterior to the Wolffian body, the **metanephros**, or permanent kidney; the duct in question being the true ureter. In the amniota the ureter grows out toward a mass of cells in which true uriniferous tubules with their associated Malpighian corpuscles are being formed. It is not certain that this structure is strictly homologous with the selachian metanephros: it may, perhaps, be so regarded, the difference in the mode of development being due to an intensified acceleration of growth similar to that already mentioned in the case of the Wolffian body. The latter structure never acts as a kidney in the higher vertebrates, its sole function being connected with the transmission of the reproductive elements of the male: the Wolffian duct becomes the spermiduct or vas deferens: in the female the whole structure is functionless and rudimentary.

This digression into the field of comparative anatomy will not be without value if it aids in making clear the brief statement of the development and homologies of the male and female urogenital apparatus of the human sub-

ject based upon the facts mentioned therein, which is
now to be entered upon.

Early in the development of the embryo there may be
seen growing downward from the dorsal wall of the body
cavity on either side a well-defined eminence, the Wolffian
or **excretory ridge**: along its base are found two ducts, the
Wolffian and Müllerian. Just mesad of the excretory ridge,
and separated from it by so shallow a groove at first as
almost to seem a modification of its mesial surface, is a
second and smaller ridge, the **germinal ridge**. The latter
becomes the gonad: the epithelium of its surface early becomes columnar and shows here and there numerous **prim-
ordial ova**; the characteristic cells being so termed without
regard to the future sex. The changes of position which
the gonads undergo in each sex will be mentioned later.

The embryonic pronephros is formed at the anterior extremity of the excretory ridge. The posterior portion, as
it developes backward, becomes the blastema or mass of
cells in which the uriniferous tubules of the kidney are
formed: it is joined by the ureter, whose extremity branches
and forms the pelvis and calyx of the kidney, and possibly
to some extent the collecting tubules: the exact relation
of the two intergrowing structures is not yet fully known.
At first the most posterior, the kidney gradually advances
in position until it occupies a location anterior to all the
associated structures.

The middle region of the excretory ridge becomes the
Wolffian body: in both sexes it develops as a series of
transverse tubules in relation with the longitudinal Wolffian duct. In the male the tubules of the anterior portion

become connected with the testis, forming the coni vasculosi of the epididymis, and possibly the rete testis as well:
those of the posterior portion, corresponding to the strictly
renal portion of the primitive kidney of the anamnia, become rudimentary and form the vasa aberrantia and the
paradidymis, or organ of Giraldes. The Wolffian duct becomes the convoluted tubule of the epididymis with its
continuation, the vas deferens, the latter giving off the seminal vesicles as it is transformed into the ejaculatory duct.
In the female the whole structure, being functionless becomes rudimentary: the anterior portion of the tubules,
corresponding to the epididymis, becomes the parovarium
or epoöphoron; the posterior portion becomes the paro-
öphoron: the Wolffian duct becomes the rudimentary duct
of Gartner.

The Müllerian duct undergoes corresponding differentiation. In the female, where it is functional, the upper portion terminates in the fimbriated extremity: the middle
portion becomes the Fallopian tube, or oviduct in the
strict sense. The lower portions coalesce on the mid-line,
the upper region of the median structure forming the
uterus and the lower the vagina: a discussion of the mode
of their coalescence would take us again into the domain
of comparative anatomy. In the male the Müllerian duct
is, like the Wolffian duct of the female, altogether functionless. It early disappears throughout the greater part of
its extent: the upper portion is perhaps represented by the
sessile hydatid of the testis: the coalesced lower portions
form the uterus masculinus, which sometimes shows traces
of a vaginal region.

In all amniota there is found very early in embryonic life a vesicular diverticulum of the ventral wall of the intestine at a point near its posterior termination: this outgrowth is called the **allantois**. In the mammals a considerable portion of it is enclosed within the body cavity: the distal portion of this becomes enlarged to form the **urinary bladder**; a narrow region corresponding to the true urethra of both sexes connects it with a proximal enlargement termed the **urogenital sinus**: the latter opens in common with the intestine into a short **cloaca** formed by a depression of the ventral surface of the body. The Müllerian and Wolffian ducts open into the sinus: the ureters at first open into the Wolffian duct, from which they are derived; later they open independently into the sinus: as growth advances they shift their position upward, until they reach their permanent point of discharge on the surface of the bladder.

The mammalian cloaca is from the first a very shallow depression: a transverse fold soon separates it into a posterior anal portion, and an anterior region which is the continuation of the urogenital sinus. Just in front of it there is found the **genital eminence**, whose posterior surface is grooved, the margins of the groove forming the **genital folds**; in front of its base the skin of the pubis is thrown into a thick fold (the future mons veneris) which is prolonged backward right and left in the **genital ridges**. In the male the genital eminence elongates; corpora cavernosa are formed in it: the genital folds coalesce from behind foward, converting the groove on the posterior surface of the eminence into a canal which is the continuation of

the narrow and elongated sinus; while the development
of erectile tissue in the walls of the canal gives rise to the
corpus spongiosum: the canal becomes the urethra of the
penis. In the female the eminence remains small and becomes the clitoris: the sinus becomes short and broad and
is represented by the vestibule: the genital folds become
the labia minores; the genital ridges the labia majores.

Glandular diverticula of the sinus are found in both
sexes. The prostate of the male may be regarded as so
derived; there is no corresponding structure in the female:
the glands of Cowper and the glands of Bartholin are homologous specializations of urethral structures. The
change of position of the gonads has been already referred to. Situated at first in the more anterior region of
the abdominal portions of the body cavity, they undergo
an apparent shifting backward in both sexes. The change
is least in the female, where they find a permanent resting
place within the pelvis. In the male they reach the brim
of the pelvis, penetrate the abdominal wall, and push their
way into the genital ridges: these enlarge, become saccular, and coalesce below the united genital fold to form
the scrotum.

CHAPTER XX.

- THE VASCULAR SYSTEM.

On account of the part played by the smaller arteries, veins and lymphatics as components of the organs of the body in which they occur, the structure of such vessels was described in one of the earlier chapters of this book. The larger vessels must themselves be regarded as organs meriting separate consideration equally with the heart, which is the central organ of the vascular system.

The **arteries** of the body differ most conspicuously from the arterioles found in the various organs by the relative increase of the media: but each of the three layers undergoes both increase and modification. The **intima** is lined, as in the smaller vessels, by endothelial cells elongated in the direction of the tube: but this endothelium rests upon a **subendothelial layer** of fibrous tissue with branched corpuscles that disappears as we pass to the finer subdivisions of the vessel: with increase in the size of the artery it becomes more fully developed. This layer is directly in contact outwardly with the elastic layer, which is well developed, especially in the larger arteries, where it has the form of a fenestrated membrane, the **membrana elastica intimae.**

The **media** consists largely of transversely disposed smooth muscular fibres, particularly in the smaller ar-

teries of the limbs: in large arteries, however, there is a distinct admixture of elastic tissue in the form of a network
of fibres; this is connected with the elastic layer of the intima, and pervades the whole muscular coat: it increases
in proportion with the size of the vessel. The muscular
fibres of the media are quite short, and are very irregular
in form, lacking the definite spindle shape characteristic
of smooth fibres in most places where they occur. In some
of the larger arteries many of the muscular bundles of the
inner part of the media are longitudinally disposed.

The **adventitia** is the stoutest and most resistant of the
coats of the arteries. It is rich in elastic tissue, especially
toward the media, the larger arteries exhibiting just external to that coat a distinct elastic layer, the **membrana
elastica externa**; this is followed by a region rich in elastic fibres: more outwardly the adventitia consists almost
wholly of closely felted bundles of white fibrous tissue
which at its outer surface passes over into the interstitial areolar tissue found between the vessel in question
and the adjacent organs. In some of the larger arteries
longitudinal bundles of smooth muscular fibres are found.

In connection with the discussion of the respiratory
tract regarded as a diverticulum of the alimentary canal,
it was pointed out that the coats of the latter were designated by characters readily demonstrable with the dis-
secting knife, but that the proper division based upon histological characters would be into an epithelial and a
musculo-skeletal layer. The coats of an artery as above
described are also examples of structures clearly distinguishable by anatomical methods: but a consideration of

the histology of the wall of an artery will show the correctness of the description that has been proposed for it as "composed of muscular and elastic tissue lined internally by endothelium and strengthened externally by a layer of connective tissue."

The features which have been mentioned as characteristic of the largest arteries are intensified in the structure of the **aorta**. In it the elastic layer of the intima is not so distinctly membranous, being composed chiefly of fibres which pass into those of the media, the two coats not being sharply defined. The media itself not only contains a very large amount of elastic tissue in proportion to the muscular tissue present, but is also reïnforced by a considerable quantity of white fibrous tissue, the coat being characterized more especially by its strength and elasticity than by its contractility. As in some of the larger arteries there are both longitudinal and transvere bundles of muscular fibres. The adventitia is not sharply defined from the media, and is relatively thin. The **pulmonary artery** agrees in most respects with the aorta in structure. The larger arteries and veins have, like other organs of the body, their own small vessels of supply; these are called the **vasa vasorum.**

While the **heart** is quite a complex organ from the anatomical standpoint, its histological structure is comparatively simple. It may be regarded as essentially a hollow mass of muscular fibres of a peculiar sort, having on its outside an investing skeletal layer, and a similar lin-

ing inwardly, each of the layers being bounded by an endothelium. The outer layer, or **epicardium,** is the cardiac
portion of the pericardium. It is invested with serous
endothelium which rests upon a membrane whose outer
portion consists chiefly of white fibrous tissue, but which
contains in its deeper layer a considerable quantity of
elastic fibres arranged in a loose and indefinite meshwork.
Beneath the membrane is a stratum of areolar tissue
which is rich in fat cells, and in which run the vessels and
nerves of the wall of the heart; in the auricles there are
numerous small ganglia connected with the nerves. The
lining membrane, or **endocardium,** is not unlike the epicardium: its investing endothelium, while continuous
with that lining the arteries and veins, and agreeing in
function with vascular endothelium, is composed of cells
which are not elongated in form, but resemble rather in
outline those found on serous surfaces. The membrane
contains a few muscular fibres, and the subjacent connective tissue but a limited quantity of fat cells. In the
auricles the elastic tissue of the membrane is quite well
developed.

The **myocardium** or muscular layer of the heart varies
greatly in thickness, being most developed in the ventricular wall and least in some portions of the auricles. The
peculiar cardiac muscular elements of which it is composed have already been described: they are arranged in
bundles which form the fibres and lamellae visible to the
naked eye: the disposition of the latter is a subject of anatomical rather than histological study. Between the
bundles is a very delicate framework of connective tissue

which supports the abundant blood and lymph capillaries: patches of adenoid tissue are also of occasional occurrence. The peculiar **fibres of Purkinje** found just beneath the endocardium of some mammals are of doubtful occurrence in man: they are large cellular elements (frequently with two nuclei) whose central portion consists of clear protoplasm but whose surface has undergone striation; they may perhaps be regarded as imperfectly developed cardiac fibres.

The **columnae carneae** and the **papillary muscles** are alike processes of the myocardium: in them the muscular fibres are chiefly disposed in a longitudinal direction, and the endocardial membrane upon their surfaces becomes almost tendinous in structure; the latter passes over from the surface of the papillary muscles into the **chordae tendineae**, whose central strands of fibrous tissue are in the larger cords reinforced by scattered bundles of muscular fibres.

The **valves of the heart** are folds of the endocardium containing bundles of fibrous tissue and a variable quantity of elastic tissue, the latter being present in greatest quantity in those places subject to the greatest pressure produced by the heart's action on the blood: the semilunar valves are farther reinforced by the presence in each of a nodule composed largely of elastic tissue, the **corpus arantii.**

The large **veins** differ from the corresponding arteries in the relative thinness of the media and the thickness of the adventitia, and in the smaller amount of elastic tissue throughout the entire structure. The endothelial cells lining the **intima** of the larger veins are but slightly if at all elongated, resembling those of the endocardium in their polygonal outline : in the smaller tributaries the form characteristic of vascular endothelium is assumed. The **media** of the two venae cavae is a continuation of the myocardium of the right auricle, and contains cardiac muscular elements : that of the larger veins generally contains but a small quantity of muscular tissue, intermingled with bundles of white fibres, which replace the elastic tissue similarly situated in the larger arteries. The media of some veins, notably those of the nervous axis and of the bones, is entirely devoid of muscular tissue. In a few cases the muscular bundles situated in the inner part of the media are longitudinal in their direction. The **adventitia** of the veins consists chiefly of a stout layer of felted bundles of fibrous tissue, containing a moderate amount of elastic fibres. In the portal, renal, and some other veins, the adventitia contains numerous bundles of smooth muscular fibres longitudinally disposed. The **valves of the veins** are folds of the intima, strengthened by bundles of fibrous tissue arranged parallel to their free borders, and containing a small amount of elastic tissue : muscular bundles from the media sometimes extend into the base.

The **lymphatic vessels** resemble the blood vessels in the fact that their walls are composed of an inner, a middle, and an outer coat: each of the three coats is, however, much thinner than in the case of the blood vessels, the middle and outer being especially so. It is customary to say that they resemble the veins in structure: it is more accurate to say that in the relative thickness of the coats and particularly in the proportion of muscular tissue they approach the arteries: they resemble the veins in the structure of the intima and in its development into valvular folds: the adventitia is relatively to the other coats weaker and simpler than in either arteries or veins. This is not the case, however, with the most highly specialized of all the lymphatic vessels, the **thoracic duct,** whose adventitia is quite stout and contains longitudinal bundles of smooth muscular fibres: the intima of the duct is also reinforced with a longitudinal network of elastic fibres. In the smaller lymphatic vessels the adventitia disappears; the media contains bundles of smooth muscular fibres obliquely disposed, in addition to the ordinary transverse bundles.

The development of the heart and of the larger vascular trunks takes place in such a manner that it is difficult to discuss it without either taking for granted a knowledge of or entering into a description of the manner in which the ground plan of the embryo body is laid down. It is therefore necessary to defer it until after the subject of Embryology has been taken up by the student.

Reference may with convenience at this time be again made to the great **serous membranes** which line the principal cavities of the body; since these latter are best re-

garded as enormous lymph cavities. Their general structure and particularly that of their characteristic epithelium has been discussed in a previous chapter: it remains
to speak briefly of some of the features characteristic of
each (in so far as they have distinguishing characteristics)
and to describe some allied though different structures
usually associated with them.

The right and left pleurae, which invest the lungs, line
the thoracic cavities and bound the mediastinum which
separates the latter, while they have the general character of serous membranes, vary somewhat in structure in
different localities. They are thickest over the ribs, where
the subserous layer of areolar tissue permits of the ready
removal of the membrane: the stomata of the endothelium are said to occur over the intercostal spaces only. The
membrane becomes thinner and more adherent as it passes
onto the surface of the diaphragm. On the surface of the
lungs it is quite thin and closely attached, the connective
tissue becoming continuous with that of the framework
of the organ: the endothelial cells of the pulmonary
pleura are also taller and more granular. The pericardium,
or heart sac, is in reality but little more than a double
serous membrane: its outer portion being derived from the
pleurae and its inner forming the proper pericardial membrane which is reflected upon the heart as the epicardium:
between the two membranes is a layer of areolar tissue
containing a small quantity of fat and numerous [blood
vessels and lymphatics.

The peritoneum, which lines the cavity of the abdomen,
is the largest and most complicated of the serous mem-

branes. Like the pleura, it is thicker in its parietal than its visceral portion. The subserous layer of areolar tissue is less developed along the ventral midline and on the under surface of the diaphragm than elsewhere in the parietal peritoneum, which is consequently most closely adherent in these regions: that of the visceral portions is in every case closely connected with the skeletal structures of the organs invested. Certain of its folds, termed **ligaments,** from their mechanical relations to the organs with which they are connected, for example, the spleen, the liver, and the uterus, have their applied fibrous membranes in close contact. Others, such as the **omenta,** have between the membranes a layer of areolar tissue in which a greater or less amount of fat may be developed. In the **mesentery** the intervening layer contains, in addition to a variable amount of fat, an extensive system of blood and lymph vessels: associated with the latter are the numerous nodules of adenoid tissue commonly called the **lacteal glands:** their structure will be described in the following chapter in connection with other similar bodies. The **tunica vaginalis** of the scrotum and testis must be regarded as an offset from the peritoneum: its structure and relations have been sufficiently described in a previous chapter.

The meninges of the brain and spinal cord, and in particular the pia mater, have been regarded as serous membranes: they differ, however, in structure and relations alike from the ordinary membranes of that name, and will best be discussed in connection with the nervous axis. The perilymphatic and endolymphatic surfaces of the internal ear will be considered in connection with the description of that organ.

The **synovial membranes,** which line the capsules of joints, the sheaths of tendons, etc., may here be considered, although they differ materially in structure from the serous membranes with which they are often associated as regards both their organization and also the characteristic fluid which fills the cavities bounded by them. They are in effect rather dense membranous layers of connective tissue, devoid of any well-defined endothelial investment: here and there cells and patches of cells may be seen upon their surfaces, some of which are sufficiently close together to assume the polygonal outline characteristic of endothelium; others, however, are distinctly branched, differing in no essential from ordinary connective tissue corpuscles: their affinity to endothelium will perhaps be evident if we recall the definition of it previously given to the effect that it is a layer of connective tissue corpuscles investing a free surface. Fringed vascular folds occur upon synovial membranes which frequently bear smaller processes, the **synovial villi,** in which a central strand of fibrous tissue is invested with a layer of small rounded cells. Articular synovial membranes pass so gradually into the fibro-cartilage which borders the true articular cartilage of the joint that it is not possible to define the boundary between the two: even the cellular elements gradually losing their processes and presenting the appearance of cartilage corpuscles.

CHAPTER XXI.

THE DUCTLESS BODIES.

The organs here included under the above title are very frequently referred to by the name of the **ductless glands.** If the term gland be used in its older and looser sense as a designation for any soft parenchymatous body, then a distinction between such of those bodies as are provided with and those devoid of ducts is natural and justifiable. It is better, however, to use the term in its narrower and more definite application to bodies composed essentially of acini or tubules lined with epithelial cells whose function is to secrete a specific solution or soluble substance, ordinarily removed through a definite channel of discharge. If the term gland be thus employed, there is but one (and a portion of another) of the bodies here considered to which the expression ductless gland may with any propriety be applied: as used to designate the others it is not only erroneous, but tends to give rise to misleading conceptions. The expression **ductless bodies,** while not without objections, is therefore preferable to the older title. The name of **adenoid bodies** has also been proposed, but is apt to lead to the false idea of a community of structure in all of the bodies in question.

It is quite important to note in this connection that the caption given to this chapter, unlike any other used hith-

erto, does not designate any system of organs associated
with the performance of one of the great functions or groups
of functions, or characterized by any common structure
or by similar anatomical or morphological relations: their
association under one heading is more than anything else
a matter of convenience; and the order of their consideration largely arbitrary, though not altogether so. The
lymphatic bodies, from their close structural and functional relation to the lymphatic vessels, might with equal
propriety be described in connection with the vascular system: the spleen, differing in important respects from the
bodies just named, resembles them in its close relation to
the circulatory system: the thymus has much in common
with them in its structure: to all these bodies (and to these
alone) the term adenoid bodies might with propriety be
applied, owing to the prevalence in them of the tissue of
that name.

The thymus, in addition to the adenoid tissue of which
it is largely composed, contains definite masses of epithelial elements whose derivation will be discussed when
that body is described: the thyroid is largely .composed
of epithelial cells which line acini; to it more than to any
other of the bodies here considered may the title of ductless gland be with propriety applied, as is shown by its
embryonic development. Resembling the thymus and
the thyroid in the origin of some of their elements (and in
little else) are the parathyroids, situated on either side of
each lobe of the thyroid; and the carotid glands; they
are closely resembled in structure, though not in origin,
by the coccygeal gland; the term gland is still univer-

sally applied to the last two; they have been called, from their associations, the arterial glands.

The **suprarenal capsules,** or the **adrenals,** as they are sometimes termed, are double structures, consisting in part of a mass of nervous tissue: they are resembled in this respect by the **pituitary body,** which consists in part of a mass somewhat analogous to the thyroid, and in part of an atrophied lobe of the brain: the **pineal body** is altogether derived from the modification of a portion of the brain, and might, in common with the pituitary body, be described in connection with that organ. This brief enumeration will suffice to show how heterogenous are the bodies here associated as regards both structure and anatomical relations: the function of most of them is at present entirely unknown. An account of the details of their structure will now be given.

The simplest **lymphatic bodies** are those rounded masses of adenoid tissue bounded by a more or less definite fibrous layer which have already been mentioned in connection with the description of the alimentary canal. The **lymphatic follicles,** as they are commonly called, of the intestine, whether solitary or clustered (as in the Peyer's patches of the ileum), and the similar masses whose somewhat confluent aggregations form the tonsils, are examples of what may perhaps best be called **lymphatic nodules.** They consist essentially of spheroidal lumps of adenoid tissue somewhat more dense in the outer than the inner portion, but showing no division into lobes. They are pervaded by capillary networks, and may be invested

by a large lymph sinus or by a plexus of small lymphatic
vessels. It is not easy to demonstrate among the latter
distinctly efferent or afferent trunks: and from the usual
position of these bodies in mucous membranes, and from
what we now know of the important bacteriophagous
function of leucocytes upon mucous surfaces it may be
questioned whether those formed in the lymphatic nodules
do not find their chief destination there rather than in the
lymph stream.

The larger and more deeply seated lymphatic bodies
commonly termed **lymphatic glands**, together with the
altogether similar bodies found in the mesentery and called
lacteal glands, for both of which the much better name
of **lymphatic nodes** has of late years come into use, are
in reality integral parts of the lymphatic and therefore of
the circulatory system. Each is a rounded body showing
a distinct depression, the **hilum,** on one side, at which
point the blood vessels of the interior enter, and from
which one or more **efferent lymphatics** leave the organ.
The surface is invested with a stout fibrous capsule, containing scattered bundles of smooth muscular fibres in the
larger nodes: it has a proper network of blood vessels,
and into it at various points pass **afferent lympahtics**
which traverse it quite obliquely, sometimes forming
small plexuses or sinuses within it before communicating with the lymph channels within.

The interior is divided into a cortical and a medullary
portion: from the capsule stout trabeculae of fibrous tissue (sometimes containing smooth muscular fibres) pass
inward: they are frequently broad and lamellar in form

and divide the cortex imperfectly into subequal lobules
sometimes called the **cortical follicles** and compared to
the lymphatic nodules (or so-called follicles) already described, from which, however, they differ in important respects: the inner extremities of their trabeculae subdivide,
communicating by their branches with a coarse network
of fibrous tissue which forms the framework of the medullary portion.

The spaces of the network just mentioned are occupied
by another network of rounded strands of adenoid tissue, the **medullary cords,** which become continuous at
their extremities with the solid **cortical lobules** of the
same tissue situated between the trabeculae. It will be
remembered that adenoid tissue consists in effect of retiform tissue whose interstices are filled with lymphoblasts:
the medullary cords and cortical lobules of adenoid tissue
are in each case smaller than the cavities in which they are
situated, the surrounding spaces containing a coarser retiform tissue (devoid of lymphoblasts and presenting but
little resistance to the passage of fluids) which connects
the cords and lobules with the medullary and trabecular
framework. A network of passages throughout the whole
node, known as the **lymph sinuses** or **lymph channels,**
is thus formed, into which the afferent lymphatics open
after passing through the capsule, and from which the
efferent lymphatics lead.

The medullary portion of the lymphatic node extends to
the surface at the hilum, the artery of supply entering it
directly: the capillary network of the interior is situated
almost entirely in the deeper portion of the adenoid

strands and masses. The lymph channels of the medulla converge at the hilum to form a plexus, from which a single efferent trunk may lead, or, in the larger nodes, several smaller vessels which unite outside the node to form a single trunk. Pacinian bodies are of frequent occurrence in the interstitial connective tissue just without the capsule at the hilum.

The relation between the lymphatic nodes and the lymph stream, as regards the formation of lymphocytes and their transformation in the blood stream into leucocytes, has been discussed in a previous chapter. The minute islands of adenoid tissue occasionally found either on or within the walls of lymphatic vessels, and known as **perilymphatic** or **endolymphatic nodules**, may be regarded as rudimentary organs of the same kind.

The **spleen** is as closely related to the blood vascular as the lymphatic nodes to the lymphatic portion of the circulatory system: it may possibly be regarded as derived from the modification of one of the last named bodies, though differing from them greatly not only in its vascular relations, but also in its internal structure. It is interesting to note that in some of the lower mammals numerous small accessory spleen-like nodules are normally present in other regions of the body. Similar bodies are sometimes found in man in the vicinity of the principal organ, of which they may be regarded as diverticula.

The spleen is invested by a serosa derived from the peritoneum which rests upon a stout capsule of fibrous tissue which, like that of the lymphatic nodes, contains occasional smooth muscular fibres: it differs from that of the bodies last mentioned in the greater predominence of elastic fibres, making the organ highly distensible. At the hilum the capsule is continued into the spleen to form large trabeculae which branch and subdivide within, eventually becoming continuous with the branches of similar though small trabeculae which pass inward from the capsule at numerous points: the trabeculae, like the capsule, contain numerous elastic fibres, and some muscular fibres: the interior of the organ is pervaded by the large-meshed reticular framework thus produced. Continuous with this framework is a coarse retiform tissue whose fibrous network is invested with branched corpuscles: in many cases the fibrous element is quite scanty, and the reticulum consists of little more than branched corpuscles connected with each other by the tips of their branches, the so-called reticular cells of the spleen. The intervals between these cells are filled with blood which contains rather more than the usual proportion of colorless corpuscles and rather less of the colored: there are present also numerous unbranched amoeboid cells somewhat larger than colorless corpuscles, the spleen-cells: these, the reticular cells, and the plasma of the blood itself contains disintegrating colored corpuscles, and pigment granules derived therefrom: the whole constitutes the spleen pulp, a reddish brown mass to which the characteristic color of the organ is due.

The splenic artery divides into several branches just before reaching the organ: these enter at the hilum, following the stout trabecular continuations of the capsule above mentioned: within the latter they branch, their branches in some cases following the subdivisions of the trabeculae. In other cases small branches leave the trabeculae and become divided into brush-like tufts of arterioles: on emerging, their adventitia, heretofore continuous with the fibrous tissue of the framework of the organ, becomes replaced by a layer of adenoid tissue: here and there this sheath is suddenly enlarged to form spheroidal masses which may be as much as a millimetre in diameter, though usually less than half as large: they are known as the **Malpighian corpuscles** of the spleen, and are readily visible to the naked eye as whitish spots in the dark brown pulp. The adenoid tissue of the corpuscles is permeated by capillaries given off from the arterioles enclosed; it is quite loose in the centre but denser at the surface, where it passes over abruptly into the retiform tissue of the pulp. The corpuscles are found surrounding small arteries and apparently strung upon them, or upon their subdivisions, in which case they look like lateral outgrowths.

On leaving the corpuscles the arteries divide into capillaries, which, like those arising from the smaller divisions of the arteries which follow the framework of the spleen more closely, finally open into the spaces of the tissue of the pulp, the endothelial cells of the capillaries gradually becoming looser, branching, and finally passing over into the reticular cells; thus affording the only instance in the body where the blood leaves the definite vessels proper to it and circulates in the interstices of the tissues; a condition largely characteristic of the circulation of all invertebrates. The veinlets of the pulp originate in the same way that the arterioles terminate, or rather its converse:

the reticular cells passing over into branched and loosely
disposed endothelial cells which later become closely united
to form the lining of the commencing vessels. The latter
soon enter the trabeculae, where they are gathered into
larger veins: these anastomose freely within the trabeculae, finally uniting to form the few large trunks that leave
the hilum.

The **thymus** is a bilobed adenoid body situated just beneath the sternum in the upper part of the thorax and extending into the lower part of the neck in the embryo and
the infant: it is gradually reduced to a mere vestige in the
adult. The whole organ is invested by a thin capsule of
fibrous tissue, beneath which it is subdivided into a number of irregular lobules each but a few millimetres in diameter. The fibrous tissue envelope of each lobule gives
off trabeculae which penetrate the interior in the same
manner as the similar structures in a lymphatic node. The
lobule in consequence exhibits a cortical and a medullary
portion. The **cortex** is composed of **nodules** of adenoid
tissue not unlike those found in a tonsil: the **medulla** is a
mass of adenoid tissue much less dense than that of the
cortex, the transition from the one to the other being so
rapid as to be quite conspicuous in sections. The trabeculae are continuous with the retiform tissue of the cortex
and medulla alike, but there is no distinct medullary framework of fibrous tissue and no segregation of the adenoid
tissue in medullary cords: nor are there any lymph channels in either cortex or medulla.

The most characteristic feature of the thymus is the

presence in the medulla of what are known as **Hassall's corpuscles,** or, as that histologist termed them, **concentric corpuscles.** These are peculiar nest-like groups of epithelioid cells which are now known to be derived from the breaking up of right and left tubular diverticula from the cervical hypoblast. Each corpuscle consists of a central granular mass containing one or more spheroidal cells, surrounded by two or three layers of concentric flattened cells: compound corpuscles sometimes occur, two or three ordinary corpuscles being invested by a common layer of concentric cells. The adenoid tissue of the organ contains a rich capillary network, and is the place of origin of numerous large lymphatics.

The **thyroid** resembles the thymus in being relatively large in foetal life and infancy: it differs from that organ in its persistence and evident functional importance in the adult, as indicated by the grave consequences of its complete extirpation. Like the tubular structures which eventually break up into the concentric corpuscles of the thymus, its characteristic elements are derived from the cervical hypoblast. It is at first provided with a duct, whose rudiment becomes the foramen caecum of the dorsum of the tongue: the duct in question in rare instances persists; in the great majority of cases it aborts, converting the organ into a true ductless gland.

Unlike most glandular bodies, the thyroid is not provided with a well-defined capsule. It is invested by a layer of areolar tissue considerably denser than that connecting it with adjacent organs but not passing over into a distinct fibrous membrane. Areolar tissue of variable density pervades the interior of the organ, forming the supporting framework of its structure, the characteristic feature of which is the presence of great numbers of **vesicles** united into imperfect lobules. The vesicles are spheroidal, polyhedral, or sometimes tubular in form, their walls consisting of a single layer of cuboidal epithelium. The interior of the vesicle is filled with a glairy yellowish **colloid substance** which frequently contains leucocytes and detached epithelial cells. A distinction has been made by some observers between the **colloid cells**, which are actively engaged in secreting the fluid contained in the vesicles, and the **reserve cells**. A definite basement membrane cannot be clearly discerned, the epithelial cells appearing to rest directly upon the interstitial septa of areolar tissue already mentioned: the areolae of the septa not infrequently contain the colloid secretion of the cells: elements resembling plasma cells are found in the interstitial tissue. The thyroid is highly vascular, the arteries being relatively quite large, and anastomosing freely: the vesicles are surrounded by a rich capillary network: the lymphatics are also large and numerous, and the presence of colloid substance in their interior may sometimes be detected.

Imbedded in the substance of the thyroid upon both the lateral and the mesial surfaces of the lobes are small bodies a few millimeters in diameter to which the name of **parathyroids** has been given. They resemble the thyroid in color and appearance, but differ from it in structure,

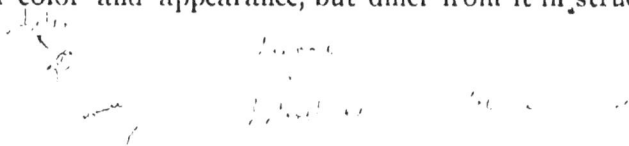

consisting of solid strands of epithelioid cells apparently anastomosing, their interspaces being occupied by numerous blood vessels. They have been regarded by some observers as masses of embryonic thyroid tissue: this view is denied by others, who regard their structure as more nearly approaching that of the carotid glands. There is usually associated with each a small mass of adenoid tissue containing concentric corpuscles and in other respects resembling the substance of the thymus.

The **carotid glands**, situated in the angle between the branches of the common carotid artery, are small irregularly shaped bodies whose envelope of connective tissue is continued inward to form a supporting framework imbedded in which are nodular masses of epithelioid cells richly supplied with capillaries. They resemble the parathyroids in their origin from the cervical hypoblast, and both are probably to be regarded as rudiments of larger and more important organs.

The **coccygeal gland** is another body of quite similar structure to those just described, and probably also rudimentary in its character. The epithelioid cells which constitute its distinguishing feature are to some extent disposed in columnar strands as in the parathyroids. Eberth has described among them nests of cells resembling concentric corpuscles. Attempts have been made to show that the elements of this body are largely nervous in character, but this view of their nature lacks confirmation. The mode of development is not known.

The **suprarenal capsules**, or, in the language of comparative anatomy, the **adrenal bodies**, since their position is usually near but not upon the kidneys, as is the case in man, are in some respects the most complex in structure of all the ductless bodies, to no other of which are they nearly allied. Each on section shows to the naked eye a distinct yellowish cortex, radially striated, and a dark brownish homogeneous medulla, the two being clearly defined from each other. The surface is invested by a thin but firm fibrous capsule whose deeper portion shows scattered bundles of smooth muscular fibres: from it tough fibrous septa enter the interior of the organ to form the cortical framework, which is limited internally by a continuous layer of connective tissue which bounds the medulla. The interior of the latter is also pervaded by a framework of fibrous tissue.

The **cortex** is divided by the difference in the disposition of its septa and in the consequent mode of segregation of its elements into three distinct zones which pass into each other without great abruptness: these are the thin **zona glomerulosa** just beneath the capsule, the **zona fasciculata** next within, which forms by far the greater portion of the cortex, and the **zona reticularis**, little if at all thicker than the outer zone, which lies next the medullary sheath of connective tissue. The spaces of the fibrous framework are occupied in the outer zone by rounded nodules, in the middle by columnar masses, and in the inner by a network of strands of closely packed polyhedral cells of moderate size whose protoplasm shows numerous small oil globules to which the color of the cortex is largely due.

The cells of the inner zone are darker in color, frequently containing brownish pigment: those of the outer in some animals are occasionally columnar in form, being disposed about an ill-defined lumen in the centre of the nodule.

The **medulla** contains within the fibrous stroma irreg‐ ular cords and masses of cells larger and much more loosely arranged than those of the cortex: they are de‐ void of oil globules and frequently exhibit branching pro‐ cesses. A rich plexus of non-medullated nerve fibres is present, and connected with great number of ganglion cells either scattered or clustered in groups of varying size. Numerous small ganglia are also found upon the nerves just external to the hilum.

The arteries of supply enter the surface of the capsule by numerous small branches: within, the vessels are distri‐ buted to the cortex along its framework, the capillaries not pervading the cellular masses as in the parathyroids, carotid glands, and coccygeal gland: thence they pass to the medulla, which contains a large plexus of veins whose branches unite into one at the hilum. The cortex is well supplied with lymphatics which communicate both with those of the capsule and with those of the medulla.

In the angles of the irregularly pyramidal adrenals of the human subject the cortex is folded upon itself, the me‐ dulla not extending into the fold; the two layers of the zona reticularis are, however, separated by a continuation of the connective tissue layer which surrounds the me‐ dulla. The distinctness between the cortex and the medulla is associated with an important difference in their embry‐ onic development. The two arise independently of each

other, and, indeed, remain so throughout life in some of
the fish-like vertebrates: the cortex arises as an outgrowth
from the peritoneum in close proximity to the mesonephros: the medulla is derived from an extension of the
adjacent sympathetic chain of ganglia: from its close connection with which and from its richness in nervous elements
many are inclined to regard the adrenals as essentially
portions of the nervous system.

The **pituitary body** is also known as the **hypophysis
cerebri**. It is a double structure, consisting of an anterior
and a posterior portion: it would be well if the former
term could be restricted to the first of these and the latter
to the other, since they are essentially different alike in
structure and in origin. The posterior division is in reality a downgrowth of the brain, as the second term implies: it is the rudiment in man and mammals of what is
a distinct and important lobe of the brain in the fish-like
vertebrates. The anterior is an upgrowth from the epiblast which lines the oral invagination, and is in substance an epithelial body: it is also a rudiment of what
was probably in the earlier vertebrates or their inverte-
brate ancestors an important glandular organ; its pathological relations indicate that it still has a persistent
though as yet unknown function. The contact and cohe-
sion of the two bodies is confined to the mammals: in all
other vertebrates they remain distinct.

The **anterior lobe** is larger than the posterior, and is of
a darker reddish color. It consists of spheroidal and simple or branched tubular acini; their closed cavities are lined

either by a mass of polyhedral epithelioid cells which fill
the cavity, or by a layer of true epithelial cells which surround a distinct though sometimes irregular lumen frequently filled with colloid substance similar to that found
in the thyroid: in some of the larger tubules cilia have
been observed upon the cells. Between the vesicles and
tubules is a framework of connective tissue which supports the numerous blood vessels and lymphatics and is
continuous with the fibrous capsule.

The **posterior lobe** is at first a hollow diverticulum of
the 'twixt-brain: in man and all mammals its cavity is
nearly or quite obliterated, and it becomes a small solid
mass. The nervous tissues characteristic of the inferior
lobe of the brain of the fish-like vertebrates fail to devel-
ope in the higher forms, their place being taken by an ingrowth of vessels and of bundles of fibrous tissue: within
the meshes of the latter are found numerous peculiar fusiform or stellate cells which are frequently pigmented.
Where a vestige of the cavity remains it is lined with columnar ciliated cells similar to those found lining the cavity
of the third ventricle. The relation of this portion of the
pituitary body to the brain is so close that the whole is
frequently described in connection with that organ: but
the larger and apparently more important anterior portion resembles the thyroid so much both in structure and
in origin as to justify its description among the anom-
olous ductless bodies.

The **pineal body,** otherwise known as the **epiphysis
cerebri** has already been spoken of in connection with the

pituitary body. Like the posterior portion of that organ, it is strictly a portion of the brain; and its structure should be included in a full account of that organ. Like its analogue first mentioned, however, it is in mammals altogether rudimentary and completely devoid of true nervous tissue: it has therefore no longer any histological relation with the nervous axis and may as a matter of convenience be described at this time. It consists essentially of a number of closed spheroidal or tubular acini lined or in most cases wholly filled with polyhedral epithelial cells, and supported by dense interstitial connective tissue. Among the epithelial cells, as well as on the outer surface of the body, are numerous gritty calcareous particles known as **brain-sand**: this is also found in some other portions of the brain. The pineal body in some of the lower vertebrates contains true nervous tissue and is connected with a peculiar median sense organ sometimes called the parietal or pineal eye.

Attention was called in the opening paragraphs of this book to the frequent use of the term Physiological Anatomy as a synonym for Histological Anatomy. The distinction between the two is very well illustrated by the facts in the case of the bodies described in this chapter. It was alike the hope and the expectation of the earlier histologists and physiologists that the investigation of the elemental structure of each organ of the body would throw great light upon its functions. This has proved to be the case in many instances, and physiology has been under correspondingly great obligations to histology, which it has been quick to acknowledge: so ready, indeed, that the student new to the subject is to some extent liable to loose sight of the important fact that in not a few instances no such happy result has followed upon advances in our knowledge of minute structure. While, for example, the difference in the form and relations of the cells lining the alveoli of serous and of mucous glands respectively can be correlated with the difference in the character of the fluids secreted by them, no one could have foretold, given the facts of structure, the associated difference in function; there is still less evident relation between structure and function in the case of the cardiac and pyloric glands of the stomach; and none at all that can as yet be discerned between the form and arrangement of the hepatic cells and the secretion of bile. What is true in numerous other instances is eminently so in the case of the ductless bodies: we have in them examples of quite complex organs whose histological anatomy has in each case been very carefully studied; but whose function is in every instance, save that of those which consist chiefly of adenoid tissue, almost if not altogether unknown.

CHAPTER XXII.

THE CENTRAL NERVOUS SYSTEM.

The ductless bodies afford us instances of organs exhibiting quite complex histological structure, concerning
whose functions little or nothing is known. A case not
exactly the converse of this is presented by the nervous
axis; whose physiological anatomy is exceedingly complex, but whose histological structure is far less so. For
example, experiments show that the white matter of either
side of the spinal cord is resolvable into a number of more
or less distinct tracts, none of which presents any single
structural feature or any combination of features by which
it can be demonstrated: much the same may be said of
the physiological centres that have been experimentally
located in the grey matter of the cord; in no case can
their positions be demonstrated by histological methods.
   Failure to distinguish between the provinces of physiological and of histological anatomy has led many writers upon the latter to include in their description of the
nervous axis many facts (in themselves of the highest importance) which are demonstrable only by methods that
are essentially physiological: to the confusion of the student, who is frequently led to expect that his sections will
show him more than our present histological methods at
least make possible; and being disappointed is apt to

doubt the validity of the statements which he has read. The present chapter will undertake no more than a brief description of such structural features as can with certainty be made out by histological methods; in other words, by a study of the form, relations, and groupings of the structural elements present. The consideration of the different regions of the brain will follow that of the spinal cord: and both will be preceded by a short account of the membranes which invest the nervous axis or line the cavity in which it is contained.

The **meninges,** or lining and investing membranes connected with the nervous axis, are three in number: of these the outer, known as the **dura mater,** or meninx fibrosa, lines more or less closely the spinal canal and the cavity of the skull; the inner, called the **pia mater,** or meninx vasculosa, closely invests the surface of the cord and the brain; while the third, the **arachnoid,** or meninx serosa, situated between the first and the second, is structurally connected with the latter, the two being derived from the differentiation of a single investing layer. On account of the intimate relation between these two meninges they are by some anatomists described as one, under the name of the **leptomeninx,** the outer membrane being termed the **pachymeninx**: a view which is supported by the comparat ve anatomy of these structures as well as by their embryology; it is also true, however, that the two meninges so designated are themselves formed from a single mass of indifferent tissue originally filling the space between the nervous axis and its case.

The **dura** is a thick and strong mass of white fibrous tissue, mixed with a small amount of elastic fibres; the constituent bundles are disposed chiefly in a longitudinal direction in the spinal portion of the membrane, those of the
cranial portion being variously disposed. In the latter region the dura is closely adherent to the cranial periosteum,
particularly at the base of the brain; dorsally the two
fibrous layers are less intimately united; it is, indeed, customary to speak of the dura as forming the intracranial
periosteum, and as composed of two layers; but the interpretation here indicated is more in accordance with the
facts of structure. The inner surface of the cranial dura is
invested throughout with serous endothelium; a similar
structure has been described upon the outer surface in
places where it is free from the periosteum. The spinal
dura is covered on both sides with a similar investment.
In addition to flattened connective tissue corpuscles, both
the cranial and the spinal dura contain numerous granule
cells. The cranial dura is separated from the cranial
periosteum here and there by the large **sinuses** which
receive the blood from the veins of the brain, and by
the accessory or **parasinoidal spaces**, but is not itself
highly vascular; the same is true of the spinal dura, between which and the periosteal lining of the spinal canal
is found the **internal spinal plexus**. Ossification occurs
normally in the principal folds of the cranial dura (the
falx and the tentorium) in some mammals.

The **pia** consists of two more or less clearly distinguished
layers, an inner or investing layer, closely applied to the
subjacent nervous mass, with whose connective tissue
framework it is directly continuous; and an outer or vascular stratum, rich in small blood vessels whose branches
enter the nervous tissue beneath. The whole membrane

is made up of interlacing bundles of white fibres, mixed
with occasional elastic fibres; in the spinal pia the prevail-
ing direction of the bundles is longitudinal. The outer
layer of the pia is connected with the arachnoid by
numerous trabeculae of connective tissue, which pass
across the subarachnoid space; this is quite extensive
along the spinal cord, and in some places around the
brain; a nearly continuous series of definitely arranged
strands passes from the pia at the sides of the spinal cord
across to the inner surface of the dura, traversing the
arachnoid; the groups of internally converging bundles
are found between the spinal nerves, with which they
alternate; the whole forms the **denticulate ligament** of
either side. Upon the free surface of the pia and along the
surfaces of the trabeculae and the ligaments just described are found flattened connective tissue corpuscles,
endothelioid in form and position. In the stroma of the
pia plasma and granule cells occur, and in some animals
great numbers of pigment cells.

The **arachnoid** is composed of loosely interwoven delicate
bundles of fibres, the intervening meshes containing numerous flattened corpuscles: the outer surface is invested with
a layer of serous endothelium; it is near to but very sparingly connected with the dura. The inner surface corresponds to the outer surface of the pia, with which it is
united by the loose mesh work of trabeculae above described.

The relations of the meninges to each other and to the
adjacent structures are as yet but imperfectly understood,
neither their structure nor their embryonic development
being as yet fully known. Such facts as are known concerning the latter, and such light as is afforded by comparative anatomy, appear to justify the view which

regards the subdural space as the primary cleavage of the mass of connective tissue which in the embryo and in the lowest vertebrates lies between the nervous axis and the walls of the cerebrospinal canal; the portion outside this cleft becoming the dura or pachymeninx, and that internal to it the leptomeninx or arachno-pial membrane; if this view be correct, the supradural (or so-called "epidural") and the subarachnoid spaces must be regarded as secondary. As a modification of this view it has been suggested that the arachnoid is in reality reflected over the inner surface of the dura, thus forming a continuous serous membrane which surrounds a serous cavity for which the name of the **arachnoid space** (as a substitute for the more familiar name of subdural space) has been proposed; and this interpretation has some facts in its support.

The **spinal cord**, surrounded by its meninges and supported by them, hangs freely in the spinal canal; throughout a great part of its extent it is nearly cylindrical; in the cervical and lumbar enlargements its transverse diameter is distinctly greater than the dorsoventral. It is divided into symmetrical halves by the so-called **anterior** and **posterior fissures**; the former of these, the **ventral fissure** of comparative anatomy, is a distinct furrow from a fourth to a third of the diameter of the cord in depth, containing a fold of the pia which bears an important relation to the blood supply of the cord: the latter, or **dorsal fissure**, while deeper than the former, is not a true cleft, the right and left masses being separated simply by a stout **median septum** given off from the inner surface of the pia. Other septa of a similar nature, but less extensive, are given off from the pia at various points along the sides of the cord, and contribute in some measure to its subdivision.

When viewed in cross section the cord appears to the naked eye to be made up of two differently appearing substances known from their color as the **white** and the **gray matter.** The latter is in each half of the cord almost entirely surrounded by the former, but not uniformly so; from its central portion on either side a narrow ridge projects dorsally (and slightly outward) almost to the surface of the cord; from its appearance in cross section it is known as the **posterior** or **dorsal cornu**; in a similar manner the gray matter projects ventrally and outward to form an **anterior** or **ventral cornu** which is much thicker than the dorsal, but does not approach the surface so nearly; in the thoracic portion of the cord a slightly projecting ridge extending directly outward between the dorsal and ventral cornu constitutes the so-called **lateral cornu.** The gray matter of the two sides of the cord is continued inward in the mid-region throughout the extent of the cord to form the **gray commissure,** by which the two halves are connected, and in which runs the **central canal**; in consequence of the disposition of the gray matter of the cord its figure as seen in cross section is irregularly H-shaped.

The **white matter** also exhibits certain conspicuous subdivisions, due in part to the disposition of the gray matter, in part to the mode of origin of the spinal nerves. The bundles of fibres which are gathered together at regular intervals outside the cord to form the posterior or dorsal roots of the spinal nerves enter the cord close to the dorsal cornu along a well-defined line on either side, which lies in a shallow groove. Between the dorsal cornu of either side and the median septum, which extends inward to the gray commissure, lies the **posterior** or **dorsal column** of the side in question; it is limited externally by the line of entrance of the dorsal roots, between which and the dorsal midline a well-marked septal process of the pia subdivides the column into two important tracts to be presently described. Between the dorsal and the ventral cornu of each side is situated the **lateral column**; and between the ventral cornu and the ventral fissure the **anterior** or **ventral column**. As has been already stated, the ventral cornu does not approach very closely to the surface, and the fibre-bundles which enter into the anterior or ventral roots of the spinal nerves are given off irregularly over a region of considerable width; there is therefore no clearly defined boundary either externally or internally between the lateral and the ventral columns; the whole region from the dorsal cornu to the ventral fissure is therefore sometimes spoken of as the **anterolateral** or **lateroventral column**. The ventral columns of the two sides are connected with each other dorsad of the ventral fissure by a narrow layer of white matter called the **white commissure**.

The appearances thus far described can all be readily seen, most of them with the naked eye; the histological structure of which they are the expression, in some respects exceedingly simple, is in others of such complexity that it has hitherto taxed the resources of histological technique to the utmost. It has long been known that the gray matter contains numerous multipolar nerve corpuscles of varying size and of a distinct but not sharply defined arrangement; their branching processes; non-medullated nerve fibres; and numbers of the neuroglia cells mentioned in a previous chapter; that the central canal is lined with a layer of columnar cells which are sometimes ciliated; that the white matter is composed chiefly of medullated nerve fibres, mingled with which are numerous neuroglia cells like those of the gray matter: that the pial septa already mentioned subdivide in the interior of the cord to·

form a framework along which are distributed the internal vessels of the cord, the general disposition of which
can readily be determined In brief, the cord is known to
consist of multipolar corpuscles and nerve fibres supported largely by neuroglia cells and provided with a
definite blood-supply. The complexity of the cord depends
not upon the number of kinds of tissue elements present,
no other active organ in the body of equal size and importance containing so few; but upon the arrangement of
the corpuscles and fibres, and particularly upon the relations existing between them; and it is only recently that
the latter has been at all satisfactorily determined by
means of technical processes of great efficiency and corresponding delicacy. It will be profitable to consider first
the more conspicuous and therefore longer known features of the histological anatomy of the cord, and subsequently the minuter details that have more recently been
discovered.

Beginning with the **gray matter**, the most conspicuous
and in every sense the most important elements are the
multipolar corpuscles scattered throughout it. While
they at first seem to be quite irregularly distributed, the
examination of a number of successive sections from the
same region of the cord, and their comparison with similar
series from other regions will demonstrate that the corpuscles are arranged in groups extending along the length of
the cord throughout the whole or definite portions of its extent with sufficient regularity to warrant their designation as distinct **ganglionic columns.** Of these the most
readily distinguishable on account alike of its extent and
the size of its constituent corpuscles is the **column of the
ventral cornu**, or, from the known distribution of the
axis-cylinder processes of most of its corpuscles to the ventral roots of the spinal nerves, the **motor-corpuscle col-**

umn; this may again be more or less distinctly divided in
some portions of the cord into a **mesial tract,** situated
adjacent to the ventral fissure (which probably supplies
motor fibres to the nerves of the dorsal muscles of the
trunk); a **ventrolateral tract,** contiguous to the mesial
and extending to the outer side of the cornu, (which
probably supplies fibres to the nerves of the lateral and
ventral muscles of the trunk); these two tracts, which are
confluent in the thoracic portion of the cord, are quite constant throughout its whole extent; and a third, the **dorsolateral tract,** which is situated, as its name implies, in
the outer side of the cornu and dorsad of the second above
mentioned; it is present chiefly in the cervical and lumbar
enlargements (and probably supplies fibres to the brachial
and sacral plexuses). Still another group, more centrally
located, can be distinguished in some sections; but it is
much less constant than those above described   The corpuscles of the ventral cornu are the largest found in the
cord, their diameters ranging in most instances from
sixty-five to one hundred and thirty micra.

At the base of the ventral cornu and near the gray commissure is a group of much smaller corpuscles, rarely ex-
ceeding thirty micra in diameter, the **mediocentral tract;**
it is well developed in the thoracic portions of the cord,
but much less so in the cervical and the lumbar. At the
base of the so-called lateral cornu, also in the thoracic region, there is a group of corpuscles of similar size, the
mediolateral tract, or, as it is called, the **column of the
lateral cornu.**

The dorsal cornu exhibits at its juncture with the gray
commissure a well-marked column which extends from the
level of the seventh cervical to that of the third lumbar
nerve ; throughout the thoracic region it is quite conspicuous, being roughly cylindrical in form and quite sharply
defined by the arrangement of the adjacent fibres; it was

the first column of corpuscles to be clearly recognized, and is known from its discoverer as **Clarke's column.** Stilling proposed for it the name of the **dorsal** (thoracic) **nucleus,** and for similarly placed aggregations situated respectively at the level of the third and fourth cervical nerves and the second and third sacral the names of the **cervical** and the **sacral nucleus.** The corpuscles of Clarke's column are next to those of the anterior cornu in size, being from thirty to ninety micra in diameter. Small corpuscles, from twenty to thirty micra in diameter, are scattered through the rest of the cornu with little definite arrangement; those close to the mesial margin are in some cases quite distinctly elongated parallel thereto and are known as **marginal corpuscles;** near the middle of the cornu the gray matter is broken up along the mesial side and to some extent along the lateral side also by bundles of fibres; the **reticular formations** thus produced have associated with them irregularly defined groups of corpuscles. The posterior cornu is capped dorsally by a translucent mass, rich in glia cells, known as the **gelatinous substance of Rolando;** it contains numerous small rounded elements, about fifteen micra in diameter, generally regarded as nerve corpuscles, as well as other larger elements, unquestionably nervous, some of which are marginal in form and position. Mention may also here be made of the fact that various observers have described **outlying corpuscles** scattered among the fibres of the white columns, both dorsal and lateroventral.

While the corpuscles are the most conspicuous elements of the gray matter, they form but a small portion thereof. By far the larger portion consists of a densely felted mass of fibres of various kinds; including small medullated fibres, nonmedullated fibres, axis-cylinder processes and their branches, and the dendritic subdivisions of the protoplasmic processes of the corpuscles. These were

formerly supposed to be continuous and to form the so-
called reticulum of the gray matter; but the evidence of
recent investigations by methods giving results of great
delicacy leads to the belief that no such continuity exists;
what is now generally believed to be the true relation of
the corpuscles and the fibres of the gray matter will be indicated later. To the interlacing fibres and fibrillae
already mentioned are added the numerous delicate filaments proceeding from the glia cells which with the connective tissue elements present make up what is sometimes called the **spongy substance** of the gray matter.
About the central canal there lies a translucent layer similar to that which caps the dorsal cornua, the **central
gelatinous substance.**

The white matter of the cord consists, as has already
been said, chiefly of medullated nerve fibres; these vary
greatly in size, the smallest being but one or two micra
in diameter, while the largest may have a diameter of over
twenty-five micra. The great majority of these fibres are
longitudinally disposed, the most conspicuous exception
to this being found in the white commissure, through
which fibres pass at greatly varying degrees of obliquity.
Between the medullated fibres are numerous neuroglia
cells so disposed as to form an interstitial framework between the subdivisions of the pial septa. Along the surface of the septa the neuroglia cells are gathered in great
numbers, forming a definite investment which is continuous outwardly with a well defined layer of considerable thickness which lines the inner surface of the pia.

While it is possible to recognize readily that certain
regions of the cord, presently to be indicated, contain
chiefly either larger or smaller medullated fibres, through
most portions the fibres vary so much and so irregularly
in size that no such division into tracts can be made as in

the gray matter, based on the size and arrangement of the fibres; the only conspicuous visible feature on which a subdivision can be based is the presence of the important secondary septum already mentioned as situated between the median septum and the dorsal cornu of either side; this passes obliquely toward the ventral edge of the median septum, thus dividing the dorsal column into two well defined portions, long known respectively as the column of Goll (next to the median septum) and the column of Burdach (next to the dorsal cornu). The remainder of the cord has been more or less accurately subdivided into tracts by other methods; while these tracts cannot be distinguished by any means now in our possession in sections of the normal adult cord, some knowledge of them is necessary in order to understand the meaning of some of the finer details of structure developed by recent histological researches in both the white and gray matter: a brief account of them and of the methods by which they have been determined will therefore be given.

We owe chiefly to Flechsig the discovery that the medullated nerve fibres of the different tracts of the cord attain their full development at different periods of embryonic and even of infantile life; the difference in question being apparently correlated with similar differences in the calling into activity of different powers and faculties of the nervous system: the study of the spinal cords of embryos at various stages of advancement thus revealing significant differences in structure not discernible in the cord of the adult. To Waller we are indebted for pointing out that an axis-cylinder when severed from the corpuscle of which it is a process rapidly undergoes degeneration: it follows from this that in cases of lesion of the cord, either pathological or experimental, not only single fibres but also tracts composed of fibres of a similar character will undergo degeneration in a manner definitely related to the place of lesion. Without entering into the physiological significance of the changes involved, it may be said that degeneration on the side of the lesion toward the brain is called **ascending**; and that on the opposite side of the lesion **descending**; and that the same terms are applied to the tracts in which the changes in question occur. The method of Flechsig and that based on the Wallerian law of degeneration give us results which coincide to such an extent as to determine quite clearly the limit of some of the tracts about to be mentioned; the evidence in favor of the existence of others, while generally regarded as satisfactory, is by no means as conclusive.

None of the tracts which are thus shown to exist in the white matter of the spinal cord are more clearly definable than those which pass from the pyramids of the medulla oblongata into the ventral and lateral columns of the cord. The fibres of the pyramids decussating while still in the medulla, the great majority of them enter a large compact bundle whose cross section is an irregularly triangular area lying (in the human spinal cord) between the dorsal cornu, from which it is separated by a thin layer of fibres, and the lateral surface of the cord, from which it is also separated by a layer of fibres save in the lumbar region, where it extends to the surface; it is known as the **lateral** (or the **crossed**) **pyramidal tract**. In man the fibres of the pyramid do not, as a rule, all decussate in the medulla, a small tract passing down as a thin layer on the portion of the surface of the ventral column which lies within the ventral fissure on the same side as the pyramid from which it proceeds; this tract is therefore called the **anterior** (or the **direct**) **pyramidal tract**; it does not extend beyond the middle of the thoracic region of the cord. It is probable that decussation goes on between the right and left tracts throughout their whole course, the fibres passing

through the white commissure, instead of occurring all at
once in the medulla: in some mammals the pyramidal
decussation is complete, and there is in consequence no
direct pyramidal tract: this is sometimes the case in the
human subject The existence of both pyramidal tracts
was demonstrated by Tuerck, though his name is usually
associated with the smaller and less constant of the two,
which is co nm only designated the column of Tuerck.

Far less clearly defined and less constant in position is
another descending tract, the ventrolateral, or anterior
marginal bundle of Loewenthal: it consists of a thin
layer of fibres situated upon or near the surface of the
ventral and a good portion of the lateral columns: its
fibres proceed fro n the cerebellar cortex of the same side.

In close contact with the ventrolateral descending
cerebellar tract, the fibres of the two mingling to a greater
or less extent, lies the ventrolateral ascending cerebellar
tract, or anterolateral ascending tract of Gowers: like
its companion, it is throughout the larger part of its
course a thin layer of fibres, which is situated between the
tract just mentioned and the surface of the cord: it is
thickest in its most dorsal portion, and is limited in that
direction by the crossed pyramidal tract.

Dorsad of the tract of Gowers and external to the
crossed pyramidal tract is the dorsolateral ascending
cerebellar tract, or the direct cerebellar tract of Flechsig.
It begins in the lower part of the thoracic portion of the
cord (below which level the crossed pyramidal tract comes
to the surface of the cord) and extends from there upwards, passing through the restiform body in the medulla
oblongata to reach the middle lobe of the cerebellum. Like
the pyramidal tracts and unlike those just described, the
tract of Flechsig is very clearly defined. Between its dorsal border and the line of entrance of the dorsal roots of
the spinal nerves, limited externally by the surface of the
cord, is a narrow tract, the marginal zone of Lissauer.

Attempts have been made to farther subdivide the white matter of the ventral and lateral columns; but the results on which these attempts are based are thus far so conflicting as to render the conclusions drawn therefrom quite doubtful: for the present it is best to include the whole of the territory enclosing the ventral cornu and perforated by the fibre-bundles of the ventral roots of the spinal nerves (excepting, of course, the tracts already designated) under the title of the **ventro-lateral root zone.**

The columns of Goll and of Burdach, situated in the dorsal region of the white matter, have already been described as limited structurally: they may be otherwise distinguished, according to their function, as the **dorso-median** and the **dorso-lateral ascending tracts.** Imbedded within the latter may be detected a small bundle of fibres with descending degeneration, known from its outline when seen in cross section as the **comma.**

We have seen that some of the tracts above described, both ascending and descending, are in direct relation with the brain: others are doubtless composed entirely or in large measure of fibres that begin and end in the cord itself: others, and particularly those of the dorsal tracts, are in direct relation with the spinal nerves; and these latter organs are in such close connection with the cord as to merit mention in this connection.

Each **spinal nerve** possesses, as is well known, a **dorsal** and a **ventral root.** The latter consists of efferent fibres chiefly if not solely motor in function, which arise from the axis-cylinder processes of the corpuscles of the ventral cornu, as has been stated, and pass almost directly out of the cord: they therefore make no important contributions to its columns. The dorsal root, in addition to its ganglion, which will be farther discussed in a subsequent paragraph, contains a few fibres probably motor in function: it consists chiefly of efferent or so called sensory fibres which are somewhat definitely divided into two groups, a mesial and a lateral, in each bundle; they penetrate the surface of the cord more or less obliquely and then bifurcate, giving rise to ascending and descending branches, the disposition of whose terminal and collateral arborizations will be described later. Recent researches by Cajal indicate that the fibres of the lateral group, which are slender, have their bifurcation in the marginal zone of Lissauer and the adjacent part of the lateral column; their collaterals are few and delicate ' and end in the dorsal cornu: the fibres of the mesial group, which are stouter, reach the columns of Goll and Burdach, and there bifurcate: their collaterals form by far the larger portion of those subsequently to be described as derived from the dorsal column, and in particular those which form the channels whereby those impulses are transmitted which are involved in simple reflex movements.

The white matter of the cord, therefore, consists of medullated nerve fibres which may be divided according to their origin and destination into three groups, the members of which are not structurally distinguishable: those which pass from the cord to end in the brain; those, commissural in character, which pass from one portion of the cord to end in another; and those which pass to the cord from the brain or from the spinal nerves. From the description of the nervous elements given in a preceding chapter it will be evident that each of these fibres consists essentially of the axis-cylinder process of a corpuscle, and ends in an arborization, giving off along its course one or more collaterals. It is probable, but not certain, that these collaterals have their terminal arborizations in regions of the gray matter of the cord homologous with that in which the terminal arborization of the fibre itself is situated.

Leaving out of consideration for the present the fibres

which pass from the cord to the brain, it may be said that
the terminal arborizations of fibres, whether collateral or
principal, which enter the gray matter of the cord are in
close contiguity or actual contact with the bodies or the
dendritic processes of the corpuscles of the gray matter:
the latter play the part of conductors (and not a nutritive
role merely), and connection is thus established as efficiently
as by the continuity of substance once supposed to exist.
An impulse thus transmitted calls forth the activity of the
corpuscle in question, resulting in a discharge along its
axis-cylinder process which undergoes a similar distribution.

Regarding the final arborization of a fibre as essentially
similar to those of the collaterals, and therefore to be classified with them, we shall make the first step toward a conception of the physiological anatomy of the cord by an
enquiry into the distribution of these structures as they
leave the various regions of the white matter. The following account thereof, as well as that of the corpuscles of the
gray matter to be subsequently given, is taken almost
wholly from Cajal.

Collaterals of the ventral (anterior) **column.** These are
larger than those from any other portion of the cord;
springing from the large axis cylinders which compose this
column they pass dorsad in irregular groups to be distributed within the ventral cornu and particularly about the
motor corpuscles. Some bundles pass to the mid-plane of
the cord and are distributed in the ventral cornu of the
opposite side, constituting the **ventral** (or anterior) **commissure of collaterals,** situated largely dorsad of the ventral commissure formed of axis cylinders.

Collaterals of the lateral column. These pass inward
to be distributed chiefly in the central region of the gray

matter of the same side: some, however, pass to the mid-plane, dorsad of the central canal, where they form a portion of the **dorsal** (posterior) or **gray commissure**: they are divided in it into two **bundles,** a **ventral** and a **nedian,** and are distributed in the central and to some extent the dorsal region of the opposite side

Collaterals of the dorsal (posterior) **column.** The tracts of Goll and of Burdach, and the marginal zone of Lissauer, from which these collaterals are chiefly derived, are formed in great part of the continuations of the fibres of the posterior roots of the spinal nerves. Four groups of collaterals may be distinguished.

Sensitivo-motor (or reflexo-motor) **collaterals:** these arise not only from the continuations of the fibres of the posterior root but also from the fibres themselves before their bifurcation; passing across the gray matter, they terminate in the ventral cornu of the same side.

Dorsal cornu collaterals: these, like the preceding, are very numerous: they traverse the substance of Rolando in groups to form immediately ventrad thereof and throughout the substance of the dorsal cornu a dense network composed of the intercrossing of their terminal arborizations.

Clarke's column collaterals: small bundles pass ventrally from the tract of Goll to terminate in the column of Clarke of the same side, there forming a thick network about the corpuscles of the column.

Commissural Collaterals: arising chiefly in the tract of Goll numerous small bundles pass to the most dorsal portion of the gray commissure, which they traverse, to be distributed in the dorsal cornu of the opposite side: thus forming the **dorsal bundle** of the commissure.

Thus, it will be seen, each white column gives off two

kinds of collaterals: those which furnish their terminal arborizations to the gray matter of the same side, and those, commissural in character, which are destined to ramify in the gray matter of the opposite side. In either case their relations are eventually directly with the corpuscles of the gray matter, whose disposition may now be further considered. Excepting in the substance of Rolando, where some elements of special form are found, the corpuscles of the dorsal, central, and ventral regions differ but little, save in size. The only important distinction to be noted in them pertains to the final disposition of their axis-cylinder processes: on this basis five groups or kinds of corpuscles may be distinguished: of these the first four send, as will be seen, their axis-cylinder processes out of the gray matter into the white, there to become medul· lated: they may be therefore distinguished as corpuscles with long axis-cylinder processes (corpuscles of the first class); while the axis-cylinder processes of the fifth end in the gray matter not far from their point of origin: they are therefore corpuscles with short axis-cylinder processes (corpuscles of the second class, corpuscles of Golgi). The five kinds of corpuscles are as follows:

Radical corpuscles, or corpuscles directly related to the roots of the spinal nerves. These are the motor corpuscles of the ventral cornua, and comprise the largest corpuscles of the cord: their axis-cylinder processes are thick and devoid of collaterals and pass in most cases directly through the lateroventral column to enter the ventral roots of the spinal nerves. From a few of the corpuscles the axis-cylinder processes traverse the gray matter to leave the cord by the dorsal root: they pass through the spinal ganglia, however, without entering into relation with their corpuscles, and must be regarded as in all probability motor in function.

The dendrites of these corpuscles are stout, long, and
very much branched. They may be distinguished as ven-
tro-external, dorsal, and internal (mesial); the latter
branch dichotomously in the vicinity of the ventral (anterior) commissure; some of the branches pass the mid-plane
and enter the ventral cornu of the opposite side, intercrossing with corresponding processes therefrom and forming
the **protoplasmic commissure** of Cajal. The ventro-ex-
ternal processes terminate in the lateroventral column,
and the posterior in different regions of the ventral cornu.

Commissural Corpuscles: these are smaller than those
just described, and provided with fewer and shorter dendrites. Golgi demonstrated their presence in all of the
regions of the gray matter, and that the axis-cylinder processes pass to the mid-plane which they cross ventrally (in
the white commissure) to be continued to the ventrolateral column of the opposite side. Cajal has shown that
they there undergo not a single continuation, but a T-
division: this indicates that the commissural axis-cylinder
process, on reaching the white matter of the opposite side,
divides into an ascending and a descending fibre of the
column.

Columnar corpuscles: we may thus designate the numerous medium-sized corpuscles scattered throughout the
whole of the gray matter, of which the axis-cylinder processes enter into vertical fibres of their own side. The
greater number of the corpuscles of this kind which occur
in the ventral cornu send their processes to the lateroventral column: those which are situated in the dorsal cornu
direct them toward the most dorsal portion of the lateral
column in many cases, though some of the corpuscles found
in the substance of Rolando and the internal portion of
the dorsal cornu send their processes to the dorsal column.

As regards the column of Clarke, two kinds of corpuscles
can be demonstrated: commissural corpuscles, whose
processes enter the ventral commissure; and columnar corpuscles, whose processes pass to the lateral column to become continuous with the fibres of the cerebellar tract.
This continuation takes place by two methods; by the
formation of a bend, which furnishes a single conductor,
ascending or descending; and by a T-division, which forms
two conductors or vertical branches, one ascending and
the other descending.

Pluricolumnar corpuscles: Elements are so termed by
Cajal of which the axis-cylinder process is divided while
still in the gray matter into two or more portions which
enter into as many nerve fibres belonging to different columns: thus, for example, an element of this kind may send
one fibre to the ventral column of its own side and another to that of the opposite side; in another the process
may divide into a fibre for the dorsal column and another
for the lateral or ventral, etc.

Van Gehuchten has proposed for the corpuscles here
designated commissural, columnar, and pluricolumnar
the names of heteromeral, tautomeral, and hecateromeral
corpuscles respectively.

Short process corpuscles: these, which are found chiefly
in the dorsal portion of the gray matter, have slender and
flexuous axis-cylinder processes which speedily end in arborizations situated in proximity to other and adjacent corpuscles.

The substance of Rolando merits special mention on account of the peculiarities of some of the corpuscles contained therein.   The latter belong chiefly to the columnar
and short process types, with some pluricolumnar cor-

puscles; whilst commissural corpuscles are not known with certainty to occur. Those characteristic of the region are of three principal forms disposed in as many concentric zones, passing from without inward, as follows.

The **marginal corpuscles** are large fusiform or flattened elements situated between the substance of Rolando and the dorsal column of white matter, thus forming a discontinuous layer. The dendritic processes line the surface of the dorsal column and there ramify: the axis-cylinder process, arising sometimes from the border of the corpuscles, sometimes from one of its processes, is directed ventrally across the substance of Rolando; it then changes its direction to reach the posterior portion of the lateral column, with one of the fibres of which it becomes continuous.

The **pyriform** or **fusiform corpuscles** are the smallest elements of the cord: their shape is quite variable, but forms indicated by the terms above are prevalent. It is characteristic of them that they are elongated dorsoventrally and have great numbers of crooked and intermingled dendrites, the greater portion of which arise from a ventrally directed stalk which is prolonged almost to the head of the dorsal cornu. The axis-cylinder process generally arises from the posterior portion of the corpuscle and passes either dorsally or laterally to become continuous with one of the fibres of the dorsal column.

The **stellate corpuscles** are situated nearest to the head of the dorsal cornu: they unite the substance of Rolando with that region by means of their abundant spinous dendrites. The axis-cylinder process is sometimes directed lengthwise of the cord; it then comports itself as a portion of a short process corpuscle, and appears to end in the substance of Rolando itself: at other times it is directed either mesially, to become continuous with a fibre of the column of Burdach, or laterally, to form a slender fibre of the marginal zone of Lissauer.

By a comparison of this description of the corpuscles as based upon the destination of their axis-cylinder processes with that previously given of their distribution in columnar tracts along the cord, it will be seen that these latter are in nearly every instance composed of corpuscles of varying relations, not even the columns of the ventral cornu consisting solely of so called motor corpuscles: Each of these elongated clusters may therefore be regarded as consisting of corpuscles so coördinated in function as to justify the title of nuclear or **ganglionic columns** frequently applied to them.

The **corpuscles of the spinal ganglia**, while they are situated without the cord, should always be associated with that structure in any attempt to form a complete conception of the central nervous mechanism, since their efferent axis-cylinder processes enter extensively into relation with the dorsal columns and cornua. Their spheroidal or pyriform contour has been described in a previous chapter, as has also the manner in which two medullated fibres are derived from the single pole or process borne by the body of the corpuscle. They are unique in the fact that one of these fibres puts them in communication with dendritic or receptive terminals which are far more remote in most instances than those of any other nervous element.

The **central canal** of the cord is lined with a columnar epithelial layer of **ependyma cells** which are frequently but not always ciliated. These are in the embryo continued by slender prolongations which reach to the pia, forming the primary framework of the cord: such continuations persist in the lower vertebrates, but their presence in adult birds and mammals is not yet well established. In connection with the ependyma cells, and possibly derived therefrom, there are found in the adult cord of the

higher vertebrates numerous **neuroglia cells** which are
particularly abundant in the white matter,, immediately '
beneath the pia and along the septa and blood vessels, in
the substance of Rolando, and in the central gelatinous,
substance.

The blood supply of the cord merits special mention. A,
single median **ventral spinal artery** (the anterior spinal
of human anatomy) runs along the ventral margin of
the fold, of the pia which enters the ventral fissure: while
a pair of **dorsal spinal arteries** (or posterior spinals),
are situated just ventrad of the dorsal roots of the
spinal nerves: branches of these arteries ramify in the pia
to form an extensive plexus. From the ventral trunk,
small vessels, the **central arterioles** of Ross, in number
several times as many as the vertebrae follow the fold of,
the pia into the median fissure as far as the white commis-,
sure. Here they turn alternately right and left to reach
the central region of the gray matter of either side, where
they break up into small arterioles and finally into capil-;
laries: the central arterioles are distributed chiefly to the
gray matter, though some of their divisions penetrate the
white matter, particularly of the lateral and ventral columns.

The dorsal vessels and the pial plexus give off great numbers of **peripheral arterioles** which follow the dorsal median septum or the other less prominent septa into the
cord: their terminal divisions are found not only in the
white matter but also in the outer portion of the gray
matter and throughout the dorsal cornua. Each arteriole,
whether central or peripheral, has its own proper capillary area, anastomoses between adjacent arterioles not
being known to occur.

Similarly disposed **central** and **peripheral venules**
carry the blood from the capillary networks to the irregu-

lar plexuses of veins in the pia and to the principal venous
trunks: these are two in number, a **ventral** which follows
the ventral artery more or less closely, and a **dorsal**,
which overlies the median dorsal septum and is not the
companion of any artery.

The **brain** is the modified and specialized anterior end of
the nervous axis. As we pass from the spinal cord into
the medulla oblongata, and thence through the region of
the pons, into the crura cerebri, we find that the fibrous
tracts which can be recognized in the white matter of the
cord here become subdivided and variously modified, some
of them soon disappearing as such, while others may be
traced almost to the most anterior portions of the brain;
while new fibrous tracts variously related, may be detected by the method of Waller and particularly by that
of Flechsig. The central canal expands here and there to
form the ventricular cavities of the brain; while its ependymal lining frequently coming in contact with the pial
investment through the absence of intervening nervous
tissues; is often thrown into vascular folds of greater or
less extent. The gray matter, which in the cord surrounds the central canal, now lies chiefly in its floor and
sides, and is penetrated and subdivided by the diversified
fibre tracts already referred to. The ganglionic columns,
which in the cord were continuous throughout the whole
or large portions of its structure, are now broken up into
more or less definite nuclear aggregates of corpuscles, serving as centres for specific cranial nerves: some of these
nuclei may perhaps be homologized with portions of some
of the ganglionic columns; while for others no such relation is discernible.

In addition to this central prolongation and modification of the cord, other structures appear connected there

with which, while they must unquestionably be regarded
as developments of the axial region, are of such size and
structural importance as to be properly regarded as substantially additions thereto. These, like the cord, and the
basal portions of the brain as well, are composed of more
or less definite fibre tracts and corpuscular areas, related
to each other and to the more axial structures.

Within recent years great progress has been made in the
process of unraveling this complex structure, and it is
safe to say that the general topography of the brain is
known. To attempt to sketch it, however, would take
far more space than the range of the present work con-
templates, and would take us largely into a field where,
while the method has been and must be chiefly that of the
histologist, the results belong rather to the domain of the
anatomist. Medullated fibres, axis-cylinder processes with
their collaterals, and multipolar corpuscles are much the
same in appearance and in relations throughout a large
portion of the brain as in the spinal cord: and we are concerned from a histological standpoint only with those
regions of the brain in which new forms of nervous elements appear or in which some special mode of combination is demonstrable. The chief of these are the cerebellum, the cerebral hemispheres, and the olfactory bulbs;
an account of the cerebellum and of the hemispheres will
now be given: that of the olfactory bulb will be deferred
until the sense organ with which it is connected has been
described.

The **cerebellar cortex** is a superficial layer of gray matter whose well marked folds form the laminae visible on
the surface of the organ to the naked eye: the middle of
each fold is occupied by a mass of white fibres in direct
relation to the gray matter. The latter shows to the
naked eye two distinct strata; an outer and paler known

as the **molecular layer**, and an inner, of a rusty brown color, called the **granular layer**. Between these, and partially imbedded in each, is a nearly continuous stratum of large corpuscles, the most characteristic of the cerebellar cortex, the **corpuscles of Purkinje**. They are pyriform or flask-shaped, the large extremity being directed inward: from the latter an axis-cylinder process is given off which passes through the granular layer to enter the white matter as a medullated nerve fibre; during its course through the granular layer it gives off collaterals which in many cases turn backward to enter the molecular layer.

The outer extremity of the corpuscle is prolonged for a greater or less distance, but usually soon divides into two principal branches, which rapidly and repeatedly subdivide to give rise to large numbers of dendritic processes, many of which are continued to the surface of the cortex: their surface is beset with short processes which end bluntly. The ramification is in every case almost entirely confined to a plane transverse to the lamella in which the corpuscle is situated.

The great majority of the corpuscles of the granular layer are exceedingly small, with large nuclei and very scanty surrounding protoplasm: their great numbers, and their appearance when stained with carmine or other similar stains led to the name above given for the region in which they occur. For a long time their nervous character was doubted or denied. The more recent technical methods have demonstrated it beyond question, and they are now known as **granule corpuscles**. Their bodies contain a relatively large quantity of rusty brown pigment, to which the characteristic color of the layer is due. A few protoplasmic processes are given off by each corpuscle: these branch sparingly to end in a small number of dendrites with thickened extremities. From the body of the corpuscle, or frequently from one of the processes, a

slender axis-cylinder process is given off, which passes
without collaterals into the molecular layer and there
undergoes a T-division to form two slender **tangential
fibres** whose course is always in the direction of the
lamella in which they occur and therefore at right angles
to the plane of the dendrites of the corpuscles of Purkinje,
with which they are in close contact as they pass. The
tangential fibres have been shown in some of the smaller
vertebrates to run the whole length of the lamellae in
which they are situated.

In addition to the small granule corpuscles there are
present in the inner layer, though in small numbers, other
nervous elements. They are situated near the outer limit
of the layer and are nearly as large as the corpuscles of
Purkinje, which they somewhat resemble in form. Their
outer region sends off numerous protoplasmic processes,
which branch irregularly in every direction to form large
numbers of dendrites: these project chiefly into the molecular layer, though many of them lie altogether within
the granular layer. From the inner region there is given off
a slender axis-cylinder process, which branches freely
almost from its origin, the whole giving rise to an extensive arborization-plexus which is situated entirely within
the granular layer. These corpuscles therefore belong to
Golgi's second type.

The finely dotted appearance to which the molecular
layer owes its name, seen when the lamella is cut transversely, is largely due to the cut ends of the tangential
fibres. There are present in this layer two kinds of corpuscles. In the deeper portion are seen numerous corpuscles of medium size and irregular form, whose dendritic
processes branch sparingly but extend for some distance
into the surrounding region : the axis-cylinder process
takes a course generally parallel to the surface of the cortex and gives off frequent collaterals; these pass inward

to be applied in each instance to the body of a corpuscle of Purkinje, upon which, or about the base of the axis-cylinder process, they terminate with little if any subdivision: a number of such fibrils surround each corpuscle of Purkinje, forming a nest or basket about it: the corpuscles from which they proceed are therefore known as **basket corpuscles.**

Throughout the molecular layer, but chiefly in the outer portion thereof, are found numerous **stellate corpuscles,** which though smaller in size, resemble the elements just described in their general form and in the appearance of their sparingly branched dendrites. The destination of their axis-cylinder processes is not known

The axis-cylinder processes of the corpuscles of Purkinje give rise to the only fibres of the central white matter of the lamella known to be sent inward from the cerebellar cortex. The fibres which pass out into it have been shown to terminate in two different methods. Some of them end in the granular layer, where their extremities branch sparingly, the subdivisions terminating in enlargements in such a way as to give to the whole somewhat the appearance of a tuft of moss: they have therefore been designated **mossy fibres.** The others find their way to the corpuscles of Purkinje, traverse their surfaces, and subdivide to follow the branchings of the dendritic processes, thus forming a terminal arborization which adheres thereto like a vine to a tree: they have therefore received the name of **climbing fibres.**

It will be seen from the above description of the cerebellar cortex that each corpuscle of Purkinje is in relation with three sets of discharging terminals: the tangential fibres of the granule corpuscles, the collaterals of the basket corpuscles, and the arborizations of the climbing fibres. We are at present entirely ignorant of the functional relations which are based on this structure.

The cerebral cortex has been regarded as composed of several distinct layers which differ from each other as regards the form and size of the contained corpuscles. As our knowledge has increased, the boundaries between these layers have been found to be less sharply defined than was at first supposed. Three can with certainty be distinguished: the outer, or so called molecular layer, the middle, or pyramidal layer, and the inner, or polymorphous layer.

The molecular layer, like that of the cerebellar cortex, owes its characteristic appearance largely to the cut ends of the collaterals and slender fibres which are densely interwoven in it and to their terminals. Its outermost portion contains numerous neuroglia cells, which just beneath the pia form an almost continuous stratum, as in the spinal cord: this has been described as a distinct layer of the cortex. Scattered throughout the molecular layer are numerous nervous elements, the corpuscles of Cajal, that histologist having first demonstrated their distinguishing characteristics. Two forms have been described by him, the fusiform and the stellate. The former are, as their name implies, spindle shaped, and give off from either end a polar process which runs parallel to the surface of the cortex: the two processes cannot be distinguished structurally, and each may take on the character of an axis-cylinder process: from each collaterals are given off at right angles or nearly so, which are invariably directed toward the surface of the cortex. In the second form the number of similar processes is increased to three or more. In each case the processes terminate eventually in ramifications which are turned toward the cortical surface.

Cajal has also described in the molecular layer a third form of corpuscle under the name of polygonal; these have several protoplasmic processes which end in dendrites

which may extend beyond the molecular layer to enter
that beneath it.   The axis-cylinder process may arise
either from the body of the corpuscle or from one of the
processes: its terminal subdivisions are confined to the
molecular layer.

The **pyramidal layer** is very commonly divided into two
strata, the **layer of small pyramids**, next the molecular
layer, and the **layer of large pyramids**, immediately subjacent.   The elements characteristic of these two layers
are, however, so nearly alike in everything but size, and
the transition from the one to the other is so gradual in
this respect, that they may, at least for the present, be
advantageously considered as one.

The most numerous of the elements peculiar to this layer
are the **pyramidal corpuscles**: their form is indicated by
their name.   The base of the pyramid is turned away from
the surface of the cortex, the apex being directed vertically upward and continued into a long tapering **ascending stem** which, even from the corpuscles most deeply
placed, extends nearly or quite to the molecular layer: it
terminates by subdivisions into a tuft of protoplasmic
processes; similar processes are given off at right angles
along its course; while others are given off from the body
of the corpuscle, and particularly from the angles of its
base: all these processes are probably to be regarded as
dendritic: their subdivisions are distributed chiefly to the
surrounding substance of the pyramidal layer, save those of
the apical tuft, which are largely situated in the molecular
layer.   An axis-cylinder process is given off from the base
of each pyramidal corpuscle, usually from a point near the
centre: it is always directed toward the white matter beneath the cortex: collaterals are, however, given off while
it is still in the gray matter, some of which run horizontally to terminal ramifications within the pyramidal

layer, while others bend upward at a right angle to terminate in the molecular layer.

Within the lower portion of the pyramidal layer are also
found, scattered here and there, certain elements confined
to the cerebral cortex, the **corpuscles of Martinotti**: they
are also found in the third or polymorphous layer. They
are spindle shaped, roughly pyramidal, or irregular in
form, their distinguishing characteristic being the distribution of the dendritic processes chiefly outward and downward, and an axis-cylinder process which generally arises
from the uppermost portion of the corpuscle, though sometimes from one of the ascending protoplasmic processes,
and ascends toward the molecular layer in which its ramifications are usually situated: in some instances the terminals lie wholly or in part in the uppermost portion of
the pyramidal layer.

The **polymorphous layer** is so called from the occurrence therein of elements which vary greatly in form as well
as in size: they may be ovoid, spindle shaped, pyramidal,
or polygonal. They agree, however, in the fact that their
long axes are as a rule disposed horizontally to the surface of the cortex, and that when a terminal or apical
stalk is present it is never vertically directed as in the
pyramidal layer. Some of the elements present are, as
has just been indicated, ascending corpuscles of Martinotti: others belong to the second type of Golgi, having
short axis-cylinder processes which break up into terminal
ramifications in the immediate vicinity of the corpuscle.
Still others give off long axis-cylinder processes which
bend downward to enter the white matter and become
medullated fibres of varying distribution.

The deeper portion of the polymorphous layer contains
chiefly small fusiform corpuscles, which has led to its distinction on the part of some histologists as a separate

layer: this is more clearly defined in the region of the island of Reil than elsewhere, where the stratum in question is separated from the rest of the cortex by intervening white matter, forming the layer visible to the naked eye, known as the **claustrum.** In most portions of the cortex the stratum in question is not clearly definable from the rest of the polymorphous layer.

Brief mention may perhaps be made of the composition of the **subjacent white fibrous layer,** although the various fibres and tracts are not histologically distinguishable, save to a certain extent by the methods of Waller and of Flechsig. Fibres formed by the development of medullary sheaths about axis-cylinder processes which descend from the corpuscles of the cortical gray matter may pass on downward as **projection fibres** to the basal ganglia, the hindbrain, or the spinal cord itself: other fibres pass as **association fibres** to other portions, more or less remote, of the cortex of the same hemisphere: while others still pass, chiefly by way of the corpus callosum, as **commissural fibres** to regions of the cortex of the opposite hemisphere: not unfrequently an axis-cylinder process may become a projection or an association fibre, and one or more of its collaterals an association or a commissural fibre; or the converse may occur. Still other fibres pass upward into the cortex by way of projection, association, or commissural tracts to end there, the terminal arborizations being situated either in the molecular or the upper portion of the pyramidal layer.

The neuroglia of the brain does not differ from that of the spinal cord to such an extent as to merit a detailed description in so brief an account of the organ as is here given. Much the same may be said of the ependyma, which lines the ventricles and passage ways of the brain:

like that of the central canal of the cord, it is composed
chiefly of columnar cells, whose free extremities bear for a
time at least cilia-like processes not known to be vibratile.
Mention should be made of the **plexuses** of the ventricles,
formed by infoldings of the pia and the ependyma, and
consisting chiefly of a rich network of small bloodvessels
supported by the former and invested by the latter. An
outline of the blood supply of the cord was given above:
that of the brain is far too complex and too much a matter of gross anatomy to be described here. Neither will
any account be attempted in this connection of the devel-
opmental history of the nervous axis, beyond the statement that it is formed entirely from an infolding of the
outer or epiblastic layer.

CHAPTER XXIII.

THE ORGANS OF SPECIAL SENSE.

——

As was stated in a previous chapter, there are certain organs in which specially modified receiving terminals are associated with more or less highly modified forms of epithelium and with other special structures of a skeletal character to form in each case an apparatus for the reception of a specific and clearly defined impression; their stimulation giving rise to sensations of flavor, odor, sound, or light, commonly called special, as distinguished from the more diffused and less clearly definable sensations of temperature, contact, resistance, etc., received by the more widely scattered and possibly less specialized terminals described in the chapter referred to.

In each case the special receiving apparatus involved has associated therewith other special structures, chiefly skeletal, whose function it is to render more intense or more specific the impression received: these associated structures being, equally with the nervous apparatus in question, essential factors of the organ of special sense. As in the case of the brain, a full description of these structures belongs rather to the province of anatomy than to that of histology, and would require far more space than can with propriety be given here: an account of the histological composition of the essential apparatus will in each case be given, together with mention of any characteristic features noteworthy in the tissues of the accessory parts, some previous knowledge of the anatomy of the organs in question being presumed.

The immediate organs of taste are the **taste-buds**, so called from their spheroidal form, which are situated in large numbers upon both the outer and the inner sides of the valleys which surround the circumvallate papillae; on the fungiform papillae; and particularly upon the loose folds just in front of the anterior pillars of the fauces which in man represent the more definitely circumscribed **foliate papillae** of some of the lower mammals: they are also found on the soft palate and the epiglottis, and are scattered here and there over the surface of the tongue. They are spheroidal bodies, almost completely embedded in the stratified squamous epithelium of the surface where they occur: the long axis is directed vertically or nearly so to the surface, and the outer extremity tapers slightly, on which account their form is sometimes described as flask shaped. The mass consists of a number of elongated epithelial cells, of which some are spindle shaped, or flattened, and are known as **sustentacular cells**: a layer of these completely covers the outer surface, their grouping recalling somewhat the surface segmentation of a cantelope: others are scattered irregularly throughout the interior. Between them lie other cells whose bodies, except just around the large nucleus, are slender and almost filamentous; these are the so called **gustatory cells.** The outer extremity of each ends in a ciliary process, the **taste-hair**, which projects with its fellows through a small circular opening in the squamous epithelium known as the **gustatory pore**: its inner extremity is slender, often bifurcated, and frequently more extensively branched; its subdivisions, which are sometimes varicose, reach to the base of the taste-bud.

Numerous attempts have been made to demonstrate a structural relation between the elements just described and the nerve fibres which pass to the taste-buds from the subdivisions of the glossopharyngeal nerve, but thus far

without success. This nerve, like the dorsal root of a
spinal nerve, bears a ganglion near its point of union with
the nervous axis; examination by the chromate and silver
method shows that the axis cylinders of the nerve fibres
end in the taste-buds by branching among the cells in the
interior, forming what are known as **intrabulbar ramifications**; the homology existing between the glossopharyngeal and the spinal nerves would indicate that these are to
be regarded as dendritic processes at the ends of long afferent fibres, similar to the "free endings" in the epidermis
described in the chapter on the nervous tissues. According to this view it is questionable whether the so called
gustatory cells are in reality nervous in character, and
some have regarded them and the sustentacular cells as
alike modified epithelial elements whose form and arrangement favors the stimulation of the nerve terminals. There
is, however, a close relationship between the senses of
taste and of smell; and the structure of the receiving apparatus of the latter suggests an explanation of the structure of the taste-buds in which the gustatory cell would
form the first member in a series of nervous elements.
Reference will be made to this after the organ of smell has
been described: the statements already made, however,
represent the present extent of our knowledge of the facts
in the case.

Mention should be made here of the fact that some of
the medullated nerve fibres of the subdivisions of the
glossopharyngeal nerve going to the taste-buds terminate
by free endings in the stratified epithelium immediately
surrounding those structures, forming what are termed
peribulbar ramifications: these are generally regarded as
fibres of general and not of special sensibility. The glands
of a serous type, known as the **glands of Ebner,** which
are closely associated with the taste-buds, have been de-

scribed in connection with the tongue. They are to be distinctly regarded as accessory to the apparatus of taste perception, the fluid secreted by them aiding in the solution of substances whose flavor is to be perceived by the taste-buds.

The sense of taste resembles most forms of general bodily sensation in requiring the actual contact of the object perceived : the remaining special senses resemble the thermal sense in being capable of giving knowledge of objects at a distance. Of these the first and as regards its receiving mechanism the simplest is that of smell. The accessory muscular and skeletal structures which make up the facial region known as the nose require no special description from a histological standpoint. The air passages which they enclose are lined with a mucous membrane known as the **pituitary** or the **Schneiderian membrane**: the **vestibule**, into which the nostril opens on either side, is lined with stratified squamous epithelium continuous at the margin of the nostril with the epidermis, of which it is a modification : the remainder is divided into two portions, the lower or **respiratory** and the upper or **olfactory** : the former is lined with stratified ciliated columnar epithelium, similar to that of other respiratory passages, beneath which is a highly vascular membrane which contains a considerable amount of adenoid tissue here and there gathered into distinct nodules, and numerous racemose glands, some of which are mucous and others are serous in character: large numbers of goblet cells are also distributed throughout the epithelium.

The **olfactory region** of the nasal mucosa can be distinguished with the naked eye by means of its well marked pigmentation, it being of a yellow color in man and some of the lower mammals, and of a yellowish brown in others. The fibrous layer is more highly vascular than in the respiratory region, but contains less adenoid tissue: it contains numerous glands, the **glands of Bowman**, which differ from the racemose glands above mentioned in being tubular, rarely branched, and but slightly bent or convoluted: the distal extremity is frequently the largest, the tube tapering toward the duct, which is always slender, and opens either upon the mucous surface or occasionally into a small ciliated crypt: the epithelium of the glands of Bowman is of the serous type, but the tubules resemble those of mucous glands in having a conspicuous lumen: ordinary racemose glands are also occasionally found in the olfactory region.

The epithelial layer of the olfactory region is composed chiefly of two kinds of elements. The first comprises the non-ciliated columnar **supporting cells**: these are chiefly prismatic in form throughout the greater portion of their extent, but with tapering inner extremities, and with oval nuclei situated at an approximately uniform distance from the surface: other more deeply situated epithelial cells are pyramidal in form, their bases resting on the fibrous layer; they may perhaps be regarded as immature supporting cells. Interspersed among the columnar epithelial cells are large numbers of slender elements of the second kind, whose outer extremities terminate in tufts of hair-like processes, the **olfactory hairs**, which project above the general surface: the middle portion is suddenly thickened to contain large spheroidal nuclei: these are the **olfactory cells**: their slender varicose inner portions are now known to be continuous with the medullated fibres of the nerves of smell. They must therefore be regarded as nervous elements, the thickened middle portion which contains the nucleus constituting the body of the corpuscle, and the peripheral portion a greatly reduced dendritic region consisting of a single protoplasmic process; while the proximal portion passes over into an axis-cylinder process which shortly becomes a non-medullated nerve fibre.

The olfactory nerve-fibres can be followed through the cribriform plate to their passage into the surface of the **olfactory bulbs**, whose structure may now be considered : as has been stated in a previous chapter, they constitute a distinct region of the brain ; but their histological structure is so intimately associated with their relation to the sense of smell as to make their description appropriate in this connection. Each olfactory bulb is a rounded mass at the anterior extremity of the longer or shorter **olfactory tract** (or olfactory nerve, improperly so called): the whole is an outgrowth from the hemisphere and originally contains a cavity, the **olfactory ventricle**, which is a diverticulum of the lateral ventricle : in many mammals this cavity persists throughout life; in some it persists in the bulb, that of the tract being obliterated ; in man and the Primates generally it disappears altogether in the adult.

In passing across a section of the olfactory bulb from the surface in close proximity to the cribriform plate to the ependymal lining of the olfactory ventricle a more or less distinct stratification may be observed: the number of layers distinguished by different observers varies according to the degree of subdivision recognized : Cajal designates five, distinguished by histological characteristics demonstrable with the aid of the silver chromate method, in addition to the ependymal layer, which is wanting in those forms in which the bulb is solid.

The first of these is the **superficial layer of nerve fibres**: this is a thin stratum of slender non-medullated fibres arranged in a felted mass: it is composed exclusively of the constituents of the bundles which pass through the perforations of the cribriform plate, whose origin, as we have seen, is in the axis-cylinder processes of the olfactory cells. The fibres leave the layer inwardly either singly or in small groups to enter the second stratum, or **layer of olfactory glomeruli**: the bodies whose presence distinguishes this layer have long been known to histologists–as spheroidal masses present in large numbers near the surface of the bulb: it is only recently that their structure has been at all understood: they are composed in part by the dense tufts of varicose fibrils which form the terminal arborizations of the olfactory fibres entering them from the superficial layer; in part by the similarly tufted dendritic ramifications of the extremities of processes derived from corpuscles situated in a deeper layer to be presently described: they are, therefore, the places where ingoing impulses are transmitted from the first to the second of a series of nervous elements.

The third stratum is termed by Cajal the **molecular layer**: as is the case with other structures similarly designated in various parts of the nervous axis, the finely punctate appearance which characterizes it when seen in section is due to the cut ends of numerous fibres, and of fibre-like processes from the corpuscles of the layer next adjacent. Cajal describes in addition, as peculiar to this layer, certain elongated or **fusiform corpuscles** whose peripheral extremities are continued by slender processes which run to the glomeruli and there terminate in small dendritic tufts which are subsidiary to those of the corpuscles of the layer next within: their proximal extremities give rise to axis-cylinder processes which run to the innermost layer and there bend strongly to pass toward the olfactory tracts, which they enter as medullated fibres.

The **layer of mitral corpuscles** is the fourth of the successive zones: it consists of large nervous elements chiefly disposed in a single stratum, whose general form is indicated by their title. The base of each corpuscle is directed toward the outer surface of the bulb: it gives off, in mammals usually from a point near its centre, a stout **descending process** which traverses the molecular layer to form in one of the glomeruli the important **dendritic ramification** already described as one of the essential constituents of each of those bodies. From the margin of the base are given off stout protoplasmic processes which diverge greatly, their ramifications interlacing to form a layer in which the corpuscles lie: their finer subdivisions extend obliquely into the molecular layer. The inwardly directed apex of the corpuscle gives rise to a stout axis-cylinder process which penetrates the layer next within and there bends abruptly to run backward in that layer, giving off collaterals whose terminal ramifications are in the molecular layer, and eventually to become the axis-cylinder of a medullated fibre of the olfactory tract.

The form and relations of the mitral corpuscles are subject in different vertebrates to variations in detail that are of such importance, as bearing upon their functions, as to merit description here. In mammals generally each mitral corpuscle bears but a single descending process: this may, however, divide and send branches to more than one glomerulus. In birds each mitral corpuscle gives off several descending processes to as many glomeruli: in either of these two ways a single corpuscle is put in relation with a number of the bipolar nervous elements of the olfactory mucosa. In some mammals, however, the glomeruli are relatively large, and each receives the dendritic ramifications of several descending processes from as many mitral corpuscles: in such cases a single bipolar element may transmit a stimulus to several mitral corpuscles: this latter condition obtains in the olfactory bulbs of mammals possessed of a high degree of olfactory sensibility.

Within the layer of mitral corpuscles is found the **granular layer** or **deep layer of fibres**: both of these terms being

applied to the fifth stratum as here defined. Omitting
from consideration the ependymal lining of the cavity of
the bulb, by some regarded as belonging to this layer, but
which does not differ in any essential respect from the
ependyma which everywhere lines the cavities of the nervous axis, the stratum consists almost exclusively of the
two kinds of nervous elements indicated by the titles
above given. The principal constituents are the **nervous
fibres**, which represent the axis-cylinder processes already
described as entering this layer from the subjacent corpuscles: as these pass along the length of the bulb they give
off numerous collaterals, of which some run horizontally
to end among the adjacent granules; others descend vertically to terminate by interlacing ramifications among
the lateral protoplasmic processes of the mitral corpuscles.

The name of **granule corpuscles** is used to designate
the abundant cellular elements, probably nervous in character, which are grouped in numerous clusters throughout the layer, the fibres above mentioned running in interlacing bundles among these clusters. They vary more or
less in form, but are usually provided at their inner extremities with several short, slender, rapidly branching
processes: the outer extremity bears a single stouter
process which runs to the inner surface of the molecular
layer, there to ramify among the lateral processes of the
mitral corpuscles. Other corpuscular elements, which are
more doubtfully nervous in function, have been described
in this layer.

The medullated fibres of this layer pass through the olfactory tract to enter the hemisphere and there to be distributed to their destinations in the cortex. In addition,
Cajal describes in the tract **efferent fibres** which pass into
the deep fibre layer of the bulb to end in arborizations
which are situated chiefly among the central processes of

the granule corpuscles. In those mammals in which the cavity of the bulb is obliterated in the adult, the ependymal lining is replaced by a gelatinous mass containing numerous neuroglia cells. The dorsal portion of the bulb is in mammals generally far simpler in structure than the ventral, the latter, by virtue of its position, being brought into far more intimate relations with the olfactory mucosa.

At the close of the description of the structure of the taste-buds reference was made to the close relationship of the senses of taste and of smell, and a possible resemblance in structure in the two organs was intimated. That resemblance, if it exists, is chiefly between the gustatory and the olfactory cells: the former are strikingly like the latter as regards their general form, and particularly as regards the central nucleated portion and the peripheral process. As we pass inward, however, the likeness is less evident: the olfactory cell is plainly a nervous element, being continued by an axis-cylinder process which ends by arborizations in one of the glomeruli of the olfactory bulb: if the gustatory cell is a nervous element, its inner portion must be regarded as a corresponding arborization-region, very greatly reduced, and probably discharging the impulses which it transmits upon the intrabulbar ramifications of the fibres of the glossopharyngeal nerve. Such an arrangement, if it exists, is without parallel as far as known; but would find its nearest representation in the olfactory apparatus. If, on the other hand, the gustatory cells are epithelial, and not nervous, the structure of the taste-bud approaches most nearly to that of a tactile or pressure-organ, a form of sensation having little relation to the sense of taste, which resembles that of smell (and no other sense) in that it enables us to take cognizance of stimuli that must be regarded as

essentially chemical. It should be noted that true gustatory sensations are also received on portions of the
tongue in which taste-buds have not been found, and that
these organs have been described upon surfaces other
than that of the tongue which are certainly not gustatory. We are here undoubtedly confronted with a problem whose solution depends upon discoveries yet to be
made.

The apparatus of sight is far more complex in structure
than that of smell, alike in its essential portions and in
those which are accessory thereto. The former include
the capsule, fibrous in man and in the mammals generally
throughout the larger portion of its wall (though partly
cartilaginous or bony in some vertebrates); the apparatus
of refraction, with its mechanisms of adjustment; and
the structures directly involved in the reception of light
stimuli and their conversion into nervous impulses. The
accessory parts are the protecting eyelids; the investing
membrane common to them and to the eyeball; and the
glands whose secretions maintain the proper condition of
this membrane. It will be convenient to proceed, in description, from the more external accessory parts to the
more deeply seated and more complex essential structures:
in each case considering anatomical characters only in so
far as necessary for the elucidation of the histology of the
parts in question.

The **eyelids** are essentially muscular folds of the skin,
modified chiefly upon their inner surfaces. The outer surface resembles the skin of adjacent portions of the face in

the presence of diminutive hairs, accompanied by small
sebaceous glands, and by occasional sweat glands: it is
thrown into small irregular folds. The underlying corium
is loose in texture: it also differs from that of the rest of
the skin of the face in the presence of considerable numbers
of branched pigment corpuscles. At the free margin of the
lid the surface curves inward; the corium becomes more
dense: and the hair follicles are suddenly enlarged for the
development of the long, stout, and recurved **cilia**, or **eye-
lashes.** Sebaceous glands open into the follicles of the
cilia, as do also some of the ducts of modified sweat
glands known as the **glands of Moll**, others opening
freely at the surface.

As the integument approaches the inner surface of the
lid it bends almost at a right angle at the **palpebral border,** and at the same time becomes modified in structure
to form the **palpebral conjunctiva**, which lines the surface of the lid in contact with the eyeball. Beneath the integument is situated the **orbicularis muscle**, composed of
striated fibres whose bundles run in a general way parallel to the palpebral border: the group of bundles situated just within the border fold, and separated from the
mass of the orbicularis by the follicles of the cilia, is distinguished as the **ciliary** or **marginal muscle,** or as the
muscle of Riolan. Just interior to the orbicularis muscle lies the **palpebral fascia**, a layer of fibrous tissue which
separates the tegumental from the conjunctival portion
of the lid: in the upper lid it is blended with the tendon of
the levator muscle.

The **palpebral conjunctiva** consists of a layer of stratified columnar epithelium containing scattered goblet cells,
and resting upon a definite basement membrane: and a
dense mass of fibrous tissue, the **tarsus,** or the tarsal cartilage (erroneously so-called, as it is entirely devoid of cartilage corpuscles). The stratified columnar epithelium of the

conjunctiva passes gradually at the palpebral border into the stratified squamous epithelium of the integument; the tarsus may be regarded as the continuation of the denser portion of the tegumentary corium. Upon the inner surface of the lid there can be seen with the naked eye. a number of vertical rows of apparently granular masses, of a yellowish color. These are the **tarsal** or **Meibomian glands**, compound structures of the sebaceous type which are imbedded in the tarsus; each consists of a straight or somewhat curved conducting tube or duct lined with cuboidal epithelium, into the sides of which open numerous sebaceous saccules resembling in every essential those found in connection with the hair follicles; the ducts open by minute orifices upon the margin of the lid, their mouths being lined for a short distance with stratified squamous epithelium.

Along the proximal margin of the tarsus, and partly imbedded therein are scattered branched tubular glands of the serous type, the **accessory tear glands**: they discharge their secretion upon the adjacent conjunctival surface. The conjunctival surface of this vicinity is frequently thrown into folds, chiefly involving the epithelium, whose appearance in cross section has led to their being described as glands. The connective tissue between the basement membrane of the conjunctiva and the tarsal plate contains diffuse adenoid tissue which is occasionally gathered into nodules in the human subject; in some of the lower mammals these nodules are quite numerous and well-defined.

Beyond the base of the eyelids the conjunctiva passes over upon the eyeball at the **fornix conjunctivae**. Goblet cells are more numerous here than upon the palpebral surface; the fibrous portion contains a number of distinct adenoid nodules, and a few mucous glands: inwardly, it passes over into a loose layer of subconjunctival areolar tissue which permits of a considerable amount of motion. The continuation of the conjunctiva upon the eyeball will be best described in connection with that structure. The **plica semilunaris**, a vertical curved fold at the inner angle of the eye, representing the third eyelid of many lower vertebrates, is a mere fold of conjunctiva; it contains internally in some mammals, and sometimes in man, a thin slip of hyaline cartilage: as well as a rudimentary racemose gland regarded as representing the Harderian gland generally present in the eyes of those vertebrates which have a functional third eyelid. The adjacent **caruncle** is a rounded fatty mass, with an investment agreeing with the integument in structure and containing minute hairs and modified sweat glands. The conjunctiva, which is highly sensitive, is richly supplied with the nerve terminals already described as end-bulbs.

The **lachrymal gland**, situated in the supero-lateral portion of the orbit, consists of two somewhat distinct portions, sometimes described as the superior and inferior lachrymal glands. The whole mass consists of an aggregation of compound racemose glands which open by independent ducts upon the conjunctival surface in the region of the superior fornix. The acini, which may be either simple or branched, are lined by granular cells with large spherical nuclei, agreeing in this respect with the alveoli of serous glands; from which they differ, however, by the presence in each of a distinct and sometimes a large lumen. They open into ductules lined by flattened or low columnar cells: these lead into ducts whose epithelium is distinctly columnar, and in which a second layer of small cells has been described as situated near the basement membrane.

The secretion of the lachrymal gland, after washing the surface of the eyeball, is carried away by the **lachrymal** canals, which open on the palpebral borders near their inner extremities. Each canal is lined with stratified squamous epithelium, which rests upon a fibrous layer rich in elastic fibres: external to this is a layer of striated muscular fibres which are generally disposed longitudinally. The canals discharge into the **lachrymal sac**, which is continued to the nasal cavity by the **nasal** or **lachrymal duct**. The sac and the duct are both composed of an elastic fibrous layer containing considerable adenoid tissue, and lined by a mucous membrane which is invested by columnar epithelium resembling that of the nasal cavity.

The visual capsule consists of two distinct strata; the outer, or skeletal, which is known throughout the greater portion of the eyeball as the **sclerotic**, but is transformed in front to form the transparent **cornea**; and the inner, or musculo-vascular, which is composed of the posterior **choroid**, and the anterior **iris**: the essential nervous structure of the eye lies immediately interior to the choroid. The cornea is a part of the refracting apparatus of the eye, and the iris a portion of the regulatory mechanism: but each may be conveniently described in connection with the stratum of which it is a portion.

The whitish **sclerotic** is a dense fibrous layer resembling somewhat a greatly thickened membrane; the interlacing fibre bundles are arranged chiefly in antero-posterior and in transverse directions: elastic fibres are sparingly present: the fixed corpuscles are flattened, and lie in definite lacunae of irregular form. The inner layer is rich in brownish pigment and is known as the **lamina fusca**: between it and the outer layer of the choroid are extensive lymph spaces lined with endothelium and traversed by blood-vessels and strands of connective tissue. The sclerotic is nearly twice as thick in its posterior as in its anterior portion: where the optic nerve enters the eye the sclerotic be-

comes continuous with the sheath of that cylindrical body:
the circular area enclosed is suddenly thinned, and is
pierced by a number of small openings; it is therefore designated the **lamina cribrosa**. Over the larger portion of
the sclerotic its outer surface is invested by a thin layer of
connective tissue loosely uniting it to the **capsule of Tenon**,
a membranous sac lined with endothelium and enclosing
the **space of Tenon** by means of which the eyeball is separated from the fat masses lining the orbit. In front the
sclerotic. is invested, as far as the scleral sulcus by which
it is separated from the cornea, by the **scleral conjunctiva**:
this consists chiefly of stratified squamous epithelium
resting upon a thin fibrous membrane which is connected
to the sclerotic by a scanty layer of looser connective tissue: it contains numerous end-bulbs, and, as is well known,
is extremely sensitive.

The **cornea** is readily distinguishable from the sclerotic,
not only by its transparency, but also by its greater convexity. Its outer surface is covered by a layer of stratified
squamous epithelium continuous with that of the scleral
conjunctiva: the layer is several cells deep, the outer elements being strongly flattened, but retaining their nuclei,
and the inner being digitated in a manner similar to the
prickle-cells of the epidermis. The deepest cells rest on a
thin, dense, homogeneous layer of closely felted fibres, the
membrane of Bowman, or **external limiting membrane**,
which possibly represents the fibrous portion of the conjunctiva. Beneath this membrane, and closely connected
with it is the **substantia propria** of the cornea, a mass of
the corneal tissue described in detail in a previous chapter:
at its margin it passes over into the substance of the sclerotic, of which it is presumably a modification. Internally
this mass is invested with a thin homogeneous elastic **internal limiting membrane**, otherwise known as the
membrane of Descemet: its inner surface is covered

with a layer of endothelium and bounds the **anterior chamber** of the eye. Around its margin the membrane of Descemet is continued by a number of processes to form the **pectinate ligament** by which the cornea is attached to the iris.

The **choroid** is nearly coextensive with the sclerotic. It consists of a vascular layer of fibrous tissue which is exceedingly rich in large pigment corpuscles, imparting to it a color which varies from brown to black: in the human eye it is dark brown in color. Its outermost portion, consisting entirely of pigmented connective tissue, forms a layer distinguished as the **lamina suprachoroidea**: it lies immediately within the lamina fusca of the sclerotic, from which it is largely separated, as has been already stated, by extensive lymph spaces lined with endothelium. The body of the choroid is rich in bloodvessels, which are disposed in two strata; an outer containing the arteries and veins, which are arranged in a characteristic manner, and an inner, the **capillary tunic**, or **tunic of Ruysch**: the fibrous structure between the two is a layer of connective tissue rich in elastic fibres which in some mammals is so well developed as to form a distinct layer which is visible through the capillary tunic and the retina, and is known as the **tapetum**. Within the capillary tunic is a thin transparent layer, the **membrane of Bruch**, or **vitreous membrane**.

The anterior portion of the choroid is modified by the foldings of its inner surface known as the **ciliary processes**, and by the thickening due to the presence of the layer of bundles of smooth muscular fibres termed the **ciliary muscle**: the whole region, including a marginal zone, the **ciliary ring**, in which the capillary tunic is less well developed than in the choroid generally, is sometimes designated the **ciliary body**. The ciliary processes, upwards of seventy in number in the human eye, are meridionally disposed folds which begin just anterior to the ciliary ring and rise gradually to the height of a half of a millimetre or so, to terminate abruptly at the margin of the iris: like the rest of the choroid, they are quite vascular, the vessels being imbedded in a pigmented stroma of connective tissue, and are limited internally by the vitreous membrane. Upon their surfaces are pouch-like depressions lined by the epithelial layer, presently to be described, with which the vitreous membrane is invested in this region; these have been called **ciliary glands**: their glandular function is, however, doubtful.

The **ciliary muscle** is by some regarded as a portion of the choroid, by others as interposed between it and the sclerotic. It is composed of bundles of smooth muscular fibres (of striated fibres in birds), most of which arise from the pectinate ligament at the region where the sclerotic, the cornea, and the iris come together: of these the greater number run meridionally to be inserted into the choroid, and are therefore sometimes regarded as forming a distinct muscle, the **tensor choroideae**: the remainder, known as radial bundles, run obliquely to those just described, assuming a direction which tends toward the centre of the eye, and being inserted in the ciliary processes. Other bundles, internally situated, are arranged in a more or less definite circular tract, known as the **ring-muscle of Mueller**. The ciliary muscle as a whole is triangular in cross section: it is thickest in hypermetropic eyes, due largely to an increase in the size of the ring-muscle.

The stroma of the choroid is continued forward from beyond the ciliary body to form the principal portion of the **iris**: the latter is highly vascular, but not so much so as the rest of the musculo-vascular layer: it is also somewhat different in texture, approaching more nearly to the structure of retiform tissue. It contains numerous pigment corpuscles, of different colors in different persons, on whose presence the color of the iris depends: an exception to this occurs in cases where the iris is blue: here pigment corpuscles are wanting in the stroma, the color depending entirely on the appearance of the post-iridal pigment (to be described later) as seen through the body of the iris.

The anterior surface of the stroma is somewhat condensed, and leaves a layer of endothelial corpuscles continuous at the irido-corneal angle with those which invest the membrane of Descemet. The posterior surface is formed by a homogenous layer continuous with the vitreous membrane of the choroid and the ciliary body: against it lies the layer of pigment corpuscles above referred to. Imbedded in the stroma of the iris near the pupillary margin is an angular layer of smooth muscular fibres, the **sphincter pupillae**: near the posterior surface are radiating bundles forming a thin layer which is not continuous, known as the **dilator pupillae**: its existence is questioned by some histologists, the demonstration of the scattered bundles of smooth muscular fibres, as distinguished from the adjacent bloodvessels and bundles of connective tissue, being quite difficult.

The region lying between the outer margin of the iris and the sclero-corneal sulcus is one of great importance; the iris, the choroid, the ciliary muscle, the cornea and the sclera all coming together in this vicinity. Just external to the **irido-corneal angle**, and among the fasciculi which make up the pectinate ligament, lies a loose network of trabeculae of white and elastic fibres whose interstitial cavities, imperfectly lined with endothelial cells, are known as the **spaces of Fontana**: they communicate freely with the anterior chamber of the eye and contain the same fluid. External to these and fairly within the sclerotic portion of the region is situated an annular space, irregularly flattened and in places subdivided: it is called the **canal of Schlemm**: whether it is a lymphatic

or a venous channel, and whether or not it communicates
with the spaces of Fontana are still matters of dispute.

The parts concerned with the processes of refraction by
means of which distinct images of things seen are formed
on the sensitive surface within the eye are the **cornea**, a
description of which has already been given; the **aqueous
humor**, a watery fluid in which leucocytes are occasionally
found, but containing no other tissue elements, which fills
the space between the cornea and the capsule of the
lens: the **crystalline lens**, with its capsule, by means of
which it is suspended at right angles to the eye immediately behind the iris; and the **vitreous humor**, which fills
the cavity posterior to the lens. The regulatory mechanisms are the **iris**, which, by modifying the size of the pupillary aperture, governs the amount of light which passes
through the lens and indirectly (to some extent) the sharpness of the image formed by it; and the **ciliary muscle**,
whose action modifies the convexity of the lens and thus
affects its definition: an account of these having already
been given, there remain for description the lens and the
vitreous humor, with the capsules by which they are
surrounded.

The **crystalline lens** is composed of an **epithelial layer**
and a **fibrous mass** whose components are modified epithelial corpuscles. The epithelial layer consists of cuboidal elements which form nearly the whole of the anterior surface: as they approach the equator they become
columnar, and as they reach that region become greatly
elongated and assume the shape of the long hexagonal
prismatic fibres of which the greater part of the lens is
composed: at the equator these retain their nuclei for
some time after birth, but those of the greater portion of
the mass are quickly lost.

The fibres run from the anterior to the posterior por-

tion of the mass, being so disposed that their extremities
come together in each region along radiating sutural
planes primarily three in number; the sutures of the anterior region alternate with those of the posterior region
in position; as a consequence, the fibres pass obliquely
from one region to another: the degree of obliquity is increased by the fact that the length of the fibres is such
that those which arise nearest the centre of the anterior
region terminate posteriorly nearest the equator; and
vice versa. As a consequence of this arrangement of the
fibres, stellate figures are formed which can be easily seen
in the artificially hardened lens, and traces of which have
been discovered in the living eye by the aid of the ophthal-
moscope. Hardening in alcohol reveals a tendency to
lamination, particularly in the outer portion, the laminae
peeling off in triangular patches which separate along the
sutural planes.

The **capsule of the lens** is a transparent elastic sac
which completely encloses that body. It is apparently
homogenous in structure in the adult, but is regarded by
some as composed of two laminae: an inner, cuticular in
character, formed by the activity of the epithelial elements
within while still embryonic: and an outer, composed of
fibrous tissue. It is thicker in front than behind, in relation with the greater change of curvature which takes
place in the more highly convex posterior surface in visual
adjustments. The support of the capsule and lens in place
will be described later.

The **vitreous humor**, or **vitreous body** (as it also called),
is a semi-fluid mass of extreme tranparency derived from
the modification of a quantity of gelatinous tissue by the
infiltration of lymph to such an extent that over ninety-
eight per cent. of its substance is water. Slender transparent fibres are scattered through the mass, and occasional corpuscles are found in it: these are generally of ex-

ceedingly irregular shape, and are often extensively vacuolated; they are probably modified leucocytes.   Under certain methods of hardening the vitreous body can be seen
to present evidences of a laminar structure; whether
this is real, or the result of the treatment employed, is not
yet certain.

The vitreous body is invested by a thin transparent
capsule, the **hyaloid membrane**, which is structureless
throughout the greater portion of its extent.   Opposite
the optic nerve it is reflected forward to line the slender
hyaloid canal which perforates the vitreous body, terminating in front at a point opposite the centre of the posterior surface of the crystalline lens: here the hyaloid
membrane is reflected outward from the anterior extremity of the canal to line the **patellary fossa**, on the anterior
surface of the vitreous body, which receives the convexity
of the lens and its capsule, the latter being in contact
with the hyaloid membrane.

At the margin of the patellary fossa the hyaloid membrane lining it becomes continuous with that investing
the outer surface of the vitreous body: here the membrane
is distinctly fibrous, the fibres in some cases penetrating
the gelatinous mass within.   It gives off from its outer
surface a fibrous layer which closely invests the ciliary
processes as the **zone of Zinn**, or **zonula ciliaris**: its free
portion extends beyond the ciliary processes to be inserted
upon the capsule of the lens at its equator, thus forming
the **suspensory ligament of the lens** above mentioned.
The mode of insertion of the suspensory ligament is such
as to leave a narrow circular space between the two
layers of the hyaloid, known as the **canal of Pettit**.

The structures directly involved in the reception of light
stimuli and their conversion into nervous impulses are
contained in the **retina**, a highly complex organ formed by

the modification of a direct outgrowth from the brain. This outgrowth, at first vesicular in form, is afterwards doubled upon itself in such a manner as to form a spheroidal cup composed of two layers and situated chiefly between the vitreous humor and the choroid, its basal stalk-like portion becoming the optic nerve. The fibres of the optic nerve are continued over the inner surface of the retina, in a manner to be presently described, as far as the outer or posterior extremities of the ciliary processes; the sinuous line which marks the limit of their distribution being known as the **ora serrata**. In front of this line the surface of the ciliary processes is invested by a much simpler **ciliary portion** of the retina, which is in turn continued into the **iridal portion**, or **uvea**, which lines the inner surface of the iris: the structure of these outlying portions will be best understood after a description of that of the principal portion of the retina.

When examined by the ordinary methods of hardening and staining the retina shows in transverse section a number of distinct strata or layers, tolerably uniform in structure throughout the greater portion of its extent. These are usually stated as eight in number: here, however, as in the other sense organs, our views have in recent years been greatly modified by the results of the chromate and silver method. An account of the eight layers will first be given, and the relation of their components as now understood subsequently considered.

Beginning at the inner or anterior surface of the retina, there may be discerned next to the hyaloid membrane which invests the vitreous body a delicate layer which is apparently (and only apparently) continuous: it has been designated the **internal limiting membrane**: it is in reality a mosaic formed by the thin expanded ends of supporting structural elements, the **fibres of Mueller**, which pass vertically toward (but not to) the outer surface of the retina.

Next to the internal limiting membrane, so called, is seen the first definite stratum of the retina, the **layer of nerve fibres**. These, which are non-medullated in most cases, are arranged in small bundles which form a plexiform meshwork over the inner surface, radiating from the optic nerve, or, more exactly, converging to it from all portions of the retina. This layer, like nearly all the others, is quite transparent, and is itself insensitive to light.

The layer of nerve fibres is succeeded outwardly by the **ganglionic layer**, a stratum of relatively large multipolar nerve corpuscles, either spheroidal or pyriform in shape, whose axis-cylinder processes are continuous with fibres of the preceding layer, and whose other processes ramify in the layer next beyond. The ganglionic layer varies in thickness in different portions of the retina, being in some places two or three corpuscles deep, but over the greater portion of its surface consisting of a single layer of corpuscles: toward the ora serrata these become separated from each other by considerable intervals.

Immediately beyond the ganglionic layer is situated the **inner molecular layer**: this, which is in most portions of the retina the thickest of the visible strata, is apparently composed of a granular mass, which is in reality the expression in section of the cut extremities of the ramifying processes of the corpuscular elements of the layers next adjacent.

External to the inner molecular layer is seen the **inner nuclear layer**, composed chiefly of closely aggregated bipolar and multipolar nerve corpuscles, the former predominating. The corpuscles vary greatly in size, but are, on the average, decidedly smaller than those of the ganglionic layer: the disposition of their processes will be de-

scribed later. Nucleated enlargements of the sustentacular elements, or fibres of Mueller, are also found in this
layer to some extent.

The inner nuclear layer is followed by the **outer molecular layer**, a granular stratum closely resembling in its
appearance the inner molecular layer, but differing from it
greatly in extent,being usually the thinnest of the various
layers of the retina.

Beyond the outer molecular layer, again, is seen a layer
of corpuscular elements, the so called **outer nuclear layer**,
which as ordinarily seen resembles the inner nuclear layer
as closely as do the two molecular layers: nearly all of the
constituent elements are distinctly bipolar, and, as will
presently be shown, their relations are quite different from
those of the elements of the inner of the two apparently
similar layers. The corpuscles of this layer are, further-
more, closely connected with the elements of the succeed-
ing layer in a manner which differentiates them sharply
from any other retinal elements.

The outer surface of the outer nuclear layer is sharply
defined, the sustentacular tissue of the retina here terminating so abruptly as to lead to the description of a definite
external limiting membrane: recent researches have
shown that the application of the term membrane in this
connection is even less justifiable than in the case of the
inner boundary of the retina.

External to the outer nuclear layer, and apparently
resting upon the so-called external limiting membrane, is
the **bacillary layer,or layer of rods and cones**. It consists
exclusively of the two kinds of elements designated by the
latter of these titles. Those termed **rods** are by far the
most numerous in nearly all portions of the retina: each
consists of an inner or basal portion, somewhat thicker
in the middle than at the extremities, and an outer or
terminal portion, slightly longer than the basal in man

and most mammals, which is of nearly uniform diameter throughout its entire length. The outer segment of each rod is transversely striated, and can be resolved into a number of thin disks by the aid of certain reagents: the outer portion of the basal segment is longitudinally striated. The **cones** resemble the rods in consisting of two segments: the inner of these is much stouter than the basal segments of the adjacent rods, and is thicker at the base than at the outer extremity: like the corresponding regions of the rods, the outer portion of the basal segment is longitudinally striated. Upon its free extremity is seated the outer segment, which is shorter than that of the rods, and tapers to a point: it shows transverse striations.

The layer of rods and cones was for a long time regarded as the limiting stratum of the retina, the adjacent **layer of pigment cells** being associated with the choroid, with which it is in close contact. With advancing knowledge of the embryological development of the eye, however, it has become evident that it must be regarded as retinal in nature. It consists of a single stratum of prismatic cells, hexagonal in form and so heavily loaded with pigment as to hide the large central nucleus: the outer surface of the cell, in contact with the choroid, is smooth: the inner is prolonged by numerous slender processes which extend between the rods and cones of the adjacent layer to a distance which varies in relation with the intensity of the stimulus acting upon the retina.

The layers of the retina, as above described, are now known to be the expression as seen under the more familiar methods of preparation of a system of nervous elements arranged in a manner not unlike that which is found in the other organs of special sense: their disposition may be briefly described as follows, disregarding for the present the layer of pigment cells just mentioned.

The rods and the cones of the bacillary layer are in real-

ity structurally continuous with the elements of the outer
nuclear layer in such a manner that they may with propriety be regarded as their peripheral prolongations, or as
their greatly modified dendritic portions. Each rod is
continued within the external limiting membrane by a
slender filament of greater or less extent which terminates
at the outer pole of one of the spindle shaped corpuscles of
the nuclear layer: the body of the corpuscle is transversely
striated in a characteristic manner: from its inner pole it
gives off a fine varicose filament, the homologue of the
axis-cylinder process, which extends to the outer molecular layer and there terminates in a small knob-like expansion, representing a greatly modified terminal arborization
the rod-corpuscles are situated at various levels in the outer
nuclear layer from the outer to the inner surface, their
peripheral and central filamentous prolongations varying
in length in a corresponding manner.

The base of each cone is continued beneath the external
limiting membrane by a stout strand of protoplasm
which passes almost immediately into the nucleated corpuscle, the latter being situated just within the membrane:
the corpuscle is continued inward by a stout smooth
fibre which passes directly across the outer nuclear layer
to end within the surface of the outer molecular layer by
a disk-like expansion from whose margin slender filaments
are given off, forming a rudimentary arborization.

The rod and cone elements, including both the bacillary
and the nuclear portions, as far as the outer molecular
layer, may be regarded as forming the first of the three
groups of nervous elements proper to the visual apparatus: they are frequently distinguished as the **neuro-
epithelial layer**, or the **layer of visual cells**.

The outer molecular layer, like the glomeruli of the olfactory bulb, may be regarded as chiefly made up of the
interlacement of arborizations and dendrites, the central

terminals of the rod and cone elements here coming into
relation with the peripheral terminals of the corpuscles of
the inner nuclear layer; or, as it is now frequently designated, the **layer of bipolar corpuscles**.

Each of these corpuscles is prolonged peripherally by a
filament which terminates at the outer molecular layer by
a group of dendritic processes: Cajal has shown that
those of some of them pass the outer portions of that
layer to form close tufts about the knob-like terminals of
the rod-elements; the central processes(or axis-cylinder processes) of the same corpuscles passing to the innermost
portion of the inner molecular layer: for these he has proposed the name of **rod-bipolars**. The others he has shown
to ramify extensively in the inner portion of the outer
molecular layer, in relation with the terminals of the
cone-elements; while their central processes terminate in
arborizations which are situated at various levels in the
stratified inner molecular layer: these he calls **cone-
bipolars**.

The same investigator has demonstrated in the outermost portion of the layer of bipolar corpuscles elements
varying in size, whose dendrites ramify in the outer molecular layer and whose axis-cylinder processes run for
longer or shorter distances horizontally to end in arborizations distributed in the same layer; for which reason
he calls them the **horizontal corpuscles** of the retina. In
the innermost portion of the same layer of corpuscles he
has described pear-shaped **amacrine corpuscles** of varying size, whose processes branch and ramify in the inner
molecular layer at various levels corresponding to those
indicated in connection with the terminals of the cone-
bipolars. The nature and functions of the horizontal and
amacrine corpuscles of this layer may perhaps still be regarded as matters of question; the bipolar corpuscles

clearly form the second members of the series of nervous elements involved in visual sensation.

The inner molecular layer resembles the outer (and, indeed, all the so-called molecular layers of the nervous system) in consisting chiefly of an interlacement of central and peripheral terminal filaments. It shows, as has been indicated, evidence of stratification, due to the termination, at more or less definite levels, of the central processes of the cone-bipolars, and the associated horizontal distribution of their arborizations; in relation not only with the terminals of the processes of the amacrine cells, but also with those of the dendrites of the corpuscles of the layer next within.

The **layer of ganglion corpuscles**, the third and innermost members of the visual series, is composed of elements which vary much in size: according to Cajal, the smallest corpuscles send their dendrites into the innermost stratum of the inner molecular layer: the largest to the outermost stratum: and those of intermediate size in like manner to the intervening strata; the arborizations
· of the terminals from the rod-bipolars being distributed in all cases in the innermost stratum. From each of these corpuscles an axis-cylinder process is given off which eventually becomes one of the fibres of the optic nerve, its terminal arborizations being situated in the brain. Cajal has also described in the optic nerve fibres which come from the brain and enter the retina, terminating by arborizations within the layer of bipolar corpuscles: he regards them as conveying centrifugal impulses.

The layer of bipolar corpuscles and that of ganglionic corpuscles, taken together, have been designated the **cerebral layer**, as distinguished from the neuro-epithelial layer or layer of visual cells. The cerebral and neuro-epithelial layers, taken together, are formed from the anterior (and principal) lamina of the collapsed optic vesicle, the layer of pigment cells alone representing the posterior lamina.

At the **macula lutea** the retinal layers are notably thickened, the layer of ganglionic corpuscles in particular becoming several cells deep. Passing towards the centre of the macula, the layers become rapidly thinned to form the **fovea centralis**, in which cone-elements only are present in the neuro-epithelial layer, and cone-bipolars in the cerebral layer, the central processes of the latter passing obliquely outward to enter the inner granular layer at the margin of the fovea. Where the optic nerve pierces the retina the retinal structures are of course wanting.

The characteristic retinal layers disappear at the ora serrata, the layer of visual cells first becoming absent. Over the **ciliary portion** of the retina the posterior lamina of the optic vesicle is represented by a layer of pigment cells as elsewhere: the anterior lamina by a layer of columnar cells lying between the pigment layer and the hyaloid membrane: they have large oval nuclei near their outer extremities. In the iridal portion of the retina both laminae are represented by layers of pigment cells. The masses of retinal pigment cells between the ciliary processes form the so-called ciliary glands already mentioned.

The apparatus of hearing, like that of sight, comprises a receiving and a transmitting mechanism in addition to the structure which contains the special terminal organs involved. The receiving neuro-epithelium is in some respects more complex than in any of the other organs of special sense: the accessory mechanisms are far simpler than those of the apparatus of sight.

The **pinna** consists essentially of a sheet of yellow fibro-cartilage covered by integument; funnel-shaped or vari-ously modified in mammals generally, it is in man crum-pled and comparatively rudimentary, but retains its char-acteristic structure. In the lobe of the ear the cartilage is replaced by a mass of fat. The skin upon the outer or convex surface does not in man differ materially from that of the adjacent portion of the head: that of the inner sur-face is thin and but slightly mobile upon the subjacent cartilage, and is devoid of sweat glands. The hairs of the integument of the pinna are in man very small, with rela-tively large sebaceous glands. The small intrinsic muscles which pass from certain folds of the cartilage to others, beneath the integument, are composed of slender striated fibres.

The **external auditory meatus**, in part cartilaginous and in part bony, is lined by a closely adhering tegumen-tary layer, continuous with the skin upon the inner side of the pinna, which grows thinner and simpler in struct-ure as it passes inward. The portion of the skin which invests the surface of the outer or cartilaginous portion of the tube contains fine hairs which, like those of the pinna, are accompanied by well developed sebaceous glands: the fibrous tissue subjacent to the corium contains in addi-tion numerous convoluted tubular **ceruminous glands** which greatly resemble sweat glands in form and struct-ure but are larger and more closely aggregated: they are farther characterized by their brownish color and the highly refracting fatty particles seen in their secretion. The lining of the deeper bony portion of the meatus is devoid alike of hairs and of glands.

The **membrana tympani**, which separates the external meatus from the middle ear, is composed of a fibrous lay-er invested outwardly by a continuation of the integ-

ument from the bony wall of the meatus, and inwardly
by the mucous membrane which lines the whole of the
tympanic cavity.   The fibrous layer consists chiefly of radiating bundles diverging chiefly from the point of attachment of the malleus: there are in addition circularly disposed bundles of fibres, chiefly near the margin of the
membrane, which form a so-called inner layer.   The tegumentary layer resembles that of which it is a continuation: the mucous layer consists of a thin membrane rich
in elastic fibres which bears a single layer of cuboidal epithelium whose component cells are devoid of cilia.

The **auditory ossicles** possess to some extent the character of dense bone, their thicker portions showing distinct though small Haversian systems of lamellae: definite marrow cavities exist in the interior of the principal
masses of the malleus and incus.   The articular surfaces
are in each case invested with hyaline cartilage.   The
muscles connected with the ossicles are composed of striated fibres.   Both bones and muscles are invested with
the mucous membrane which lines the tympanic cavity.

The **Eustachian tube**, which connects the tympanic
cavity with the pharynx, has a bony wall in its posterior
portion; in the anterior portion the wall is in part composed of hyaline cartilage, sparingly reinforced by bundles of white fibres, and in part of dense fibrous membrane.
It is lined throughout its extent by a mucosa which is a
continuation of that of the pharynx, and is in turn continued by the lining of the tympanic cavity: like that of
the pharynx, the mucosa is invested with a layer of stratified columnar ciliated epithelium.   In the cartilaginous
portion of the tube there is a submucosa which contains
numerous mucous glands and a considerable, quantity of
diffuse adenoid tissue; in this respect, again, recalling the
structure of the pharynx.   In the bony posterior portion

the mucosa is devoid of glands and adheres more closely to the wall of the tube.

The cavity of the **tympanum**, like that of the mastoid cells leading out from it, may be regarded as an expansion of that of the Eustachian tube. It is lined with a mucous membrane which has already been frequently referred to: between the membrane and the bony wall of the cavity is a submucosa consisting of interlaced fibrous bundles among which are seen numerous spheroidal bodies in many ways resembling Pacinian corpuscles.

The disposition of this fibrous network, the irregularities of the bony surfaces involved, and the structures present in the tympanic cavity coöperate to throw the mucosa into conspicuous folds; their disposition is a matter for the anatomist rather than for the histologist. The epithelium of the tympanic cavity is columnar ciliated over the greater portion of the surface; that of the mastoid cells is devoid of cilia. The existence of distinct glands in the mucosa of the tympanum is a matter of question.

The inner ear is in the strict sense the organ of hearing, the middle and outer regions being merely accessory thereto. It consists of the **membranous labyrinth**, in which the specially modified neuro-epithelial structures involved in audition are situated, enclosed in the cavity within the periotic mass known as the bony labyrinth. The membranous labyrinth is formed by the ingrowth of the integument of the side of the head: this is at first a simple saccular or flask-shaped cavity lined with epithelium derived from the ectoderm, and communicating with the surface by a small aperture. Later, this aperture is obliterated and the sac is divided into two principal regions, the **utricle** and the **saccule**, which are in the adult only indirectly connected. The three **semicircular canals**

are developments of the wall of the utricle and together
with it form the labyrinth in the limited sense in which the
term was formerly used. The **cochlea** is an extension of
the saccule.

The bony walls of the cavity which contains the utricle
and the semicircular canals are lined with a thin periosteum invested by flattened connective tissue corpuscles
which form an endothelium. The membranous structures
enclosed within are adherent to the periosteum along one
side of each canal and upon a portion of the surface of the
utricle: throughout the rest of their extent they are separated therefrom by a space filled with perilymph which is
traversed by frequent trabeculae of fibrous tissue, the free
surface being similarly invested. The membranous wall
consists of a layer of connective tissue containing numerous elastic fibres, within which is a dense clear **tunica
propria**, whose surface throughout the canals shows numerous low papillary eminences; lining the tunica is a
layer of polygonal pavement-epithelium cells.

In the ampulla of each of the semicircular canals the
tunica propria is much thickened along a projecting ridge,
the **transverse septum**, upon whose summit is situated a
crista acustica, or ampullar area of auditory neuro-epithelium. In passing from the general surface of the ampulla upon the sides of the septum the pavement epithelium becomes first cuboidal and then columnar in form,
the columnar cells being surmounted by a distinct cuticular layer. Within the crista the epithelium consists of
cells of two sorts: **fibre cells**, whose elongated bodies extend through the whole epithelium, their bases being
larger than their free extremities, and their nuclei being
variously situated within the basal half; and **hair-cells**,
cylindrical elements which are situated in the outer half
of the epithelium only, their nuclei being situated near

their rounded inner extremities, and their free ends bearing long tapering filaments, the **auditory hairs**.

Branches of the auditory nerve are distributed to each crista: as the fibres enter the epithelium they loose the medullary sheath, and quickly divide into fibrils which ramify extensively in the vicinity of the hair-cells, their free extremities being in every case in direct contact with these epithelial elements: the relation between the nerve terminals and the epithelial cells must be regarded as similar to and as specializations of that elsewhere described in connection with the free endings of nerve fibre in the epidermis.

In material hardened for section cutting the surface of each crista is found to be covered by a dome-shaped mass of a clear colorless substance of unknown composition and origin, in which the auditory hairs are imbedded: to this mass the term **cupula** is applied. Under suitable reagents the auditory hairs can be made to break up into numerous fine cilia-like filaments, indicating that the hairs are compound structures.

The surface of the utricle bears a large patch of neuro-epithelium, the **macula acustica** or **macula cribrosa**, as it is sometimes termed, essentially similar in structure and in the mode of nerve supply to one of the cristae of the ampullae. There is not such a marked thickening of the subjacent tunica propria, and the auditory hairs are not as long as those of the ampullar organs: the surface of the macula is invested by a soft gelatinous mass in which are imbedded numbers of crystals of calcium carbonate known as **otoliths**. A macula in every way similar to that of the utricle is found in the saccule.

The **cochlea**, a development of a portion of the saccule, with which it is directly connected in the lower vertebrates and in the embryo, is in the mammal in great measure constricted off from that region in the adult, being connected with it only by a slender tubular passage, the **canalis reuniens**. It should be regarded, however, as a tubular diverticulum of that division of the primary auditory vesicle, differing from the rest of the membranous labyrinth in its spirally coiled form, its mode of attachment, and particularly in the complexity of its neuro-epithelium, which here attains a degree of specialization found in no other organ

Regarding, for convenience, the position of the whole structure as so far shifted from that which it occupies in the living body as to bring the base of the spiral into a horizontal plane, the apex pointing upward, the cochlear tube may be said to be adherent outwardly for about one third of its surface to the bony wall of the containing cavity; and to be connected inwardly with the central bony spiral lamina by two flat membranes, the lower of which, the **basilar membrane**, is nearly horizontal in this position, while the upper, the **membrane of Reissner**, slopes at an angle of about forty-five degrees: the tube is therefore approximately triangular in cross section.

The periosteum of the outer wall is much thickened along the area of adhesion of the cochlear tube to form the **spiral ligament**, the greatest elevation being at the point of attachment of the basilar membrane, where a fibrous ridge is found known as the **crista basilaris**; a short distance above this a second ridge is seen, the **vascular prominence**, containing one or more conspicuous bloodvessels: the somewhat concave surface between this and the ridge to which the membrane of Reissner is attached, known as the **stria vascularis**, exhibits a histological structure without parallel in the entire body. It contains a rich network of capillaries, imbedded in elements apparently epithelial in character, and commonly so described: the superficial cells (which entirely overlie the capillaries) are certainly in continuity with the epithelium lining the rest of the tube; those situated between the capillaries may very possibly be epithelioid connective tissue corpuscles not unlike those known to occur elsewhere.

The **membrane of Reissner** is an exceedingly delicate sheet of connective tissue invested on the side toward the scala vestibuli with a layer of endothelium some of whose cells are pigmented: the inner side is lined, like the greater portion of the membranous labyrinth, by a pavement epithelium composed of polyhedral cells: the three component layers are of nearly equal thickness. The inner and lower edge of the membrane is united to the middle or inner portion of the **limbus**, a peculiar thickening of the periosteum of the upper surface of the bony spiral lamina.

The portion of the limbus situated outwardly from the attachment of the membrane of Reissner terminates abruptly by a border excavated by the **spiral groove**, which is bounded by an upper and a lower lip. The upper surface of the upper lip is ridged and grooved and its margin developed into numerous tongue-like processes, the **auditory teeth**: the surface of the ridges and the teeth is invested by polyhedral pavement epithelium: that of the grooves is columnar: this is continued over the spiral groove by a layer of cuboidal cells continuous at the lower lip of the groove with the epithelial structures upon the upper surface of the basilar membrane. The lower lip extends to the margin of the bony spiral lamina.

The **basilar membrane** extends from this margin to the basilar crest of the spiral ligament. Its middle layer consists of a sheet of homogeneous ground substance containing scattered nuclei, and having embedded in it an immense number of straight stout fibres running radially from the spiral lamina to the basilar crest. The surface toward the scala tympani is covered by a layer of connective tis-

sue whose elements do not take on a definite endothelial
form, but are largely spindle-shaped fibres disposed at
right angles to the fibres of the middle layer. The inner
surface is invested with epithelium continuous with that
lining the rest of the tube: that of the outer half of the
membrane, or **zona pectinata**, is but slightly modified:
that of the inner half, or **zona tecta**, is greatly modified
to form the characteristic structure of the cochlea.

This structure, **the organ of Corti**, is a neuro-epithelium
not unlike those found in the cristae and maculae acusticae:
being composed, like those bodies, of hair cells and supporting cells: its greater complexity is due chiefly to the
form and arrangement of the latter elements. The central
feature is a series of arches formed by the convergence
above of an inner and an outer **rod of Corti**, the outer
rods being longer and more slanting than the inner: the
triangular space beneath them, which runs the whole
length of the cochlea, being known as the **tunnel of Corti**.
Each rod consists of a broad basal portion, or foot, a
slender shaft, and an enlarged head, that of the inner rod
having a concave surface upon its outer side into which is
fitted a corresponding convexity upon the inner aspect of
the head of the outer rod: both the inner and the outer
rods bear outwardly directed flattened **phalangeal processes**, those of the inner rods overlapping the inner portions of the processes of the outer rods. Both inner and
outer rods are invested with a layer of protoplasm which
is accumulated at the base on the side toward the tunnel
in a mass containing an oval nucleus. The inner rods are
narrower and more numerous than the outer.

On the inner side of the upper extremities of the inner
rods is situated a row of **inner hair cells**, cylindrical in
form, and, like those of the auditory structures, only
extending through the upper half of the layer: the upper
extremity of each bears a number of hair-like processes:

the lower extremity is rounded and contains a spherical nucleus. Internal to the inner hair cells are columnar supporting cells which pass gradually over into the epithelium of the spiral groove.

On the outer side of the heads of the outer rods are rows, three or four in number, of **outer hair cells**, essentially like those of the inner row. Between them are the upper extremities of the **cells of Deiters**, elements somewhat resembling the rods of Corti: each has a spindle-shaped basal portion containing a spheroidal nucleus, and a slender rigid upper portion which terminates in an outwardly directed **phalanx**. The phalangeal processes of the rods of Corti and the phalanges of the cells of Deiters are united by their angles to form a **reticular membrane** through whose apertures the extremities of the outer hair-cells project. Between the outer rods, the hair-cells, and the cells of Deiters are intervals, the **spaces of Nuel**, which communicate with each other and with the tunnel of Corti, the whole being filled with a semifluid substance.

The organ of Corti may be said to be limited by the hair cells and the cells of Deiters: the latter pass over outwardly into tall columnar elements, the **cells of Hensen**, whose nuclei are situated in their large upper extremities. These pass rather abruptly into the shorter columnar **cells of Claudius**, between which and the epithelium of the outer wall of the tube a gradual transition is seen along the surface of the zona pectinata.

The bundles of nerve fibres distributed along the cochlea pass along the under surface of the bony spiral lamina to its margin from the spiral ganglion of the modiolus. Here they penetrate the basilar membrane: the fibres loose their medullary sheath and are distributed to the epithelium in a manner quite similar to that described in the account of the cristae, some of the fibres traversing the tunnel of Corti to reach the vicinity of the outer hair cells. From the margin of the upper lip of the limbus a cuticular fold, the **membrana tectoria**, extends out as far as the outer cells. It probably rests upon the organ of Corti during life.

A comparison of the essential structures of the various organs of special sense shows that they agree in being modified epithelia containing more or less specialized terminals of nervous elements. These epithelia are in each instance derived from the ectoderm: in the case of the retina indirectly, the organ in question being formed as a diverticulum of the nervous axis, which is itself formed from an infolding of the ectoderm: in the case of the other sense organs the derivation from .the ectoderm is direct. The nerve terminals of the organs of taste and of hearing resemble each other in consisting of fibrils ramifying between the specialized epithelial cells, though these senses are not at all related as regards the character of the stimuli to which they respond. Similarly, the terminals of the organ of smell and of sight are somewhat alike, greatly as these senses differ. Farther investigation may explain these apparent resemblances and differences and exhibit a still deeper unity of structure in the mechanisms of special sense.

INDEX.

www.ingramcontent.com/pod-product-compliance
Lightning Source LLC
Chambersburg PA
CBHW021400210326
41599CB00011B/948